A New Name for Peace

A New Name for Peace

International Environmentalism, Sustainable Development, and Democracy

Philip Shabecoff

University Press of New England
Hanover and London

University Press of New England, Hanover, NH 03755

Printed in the United States of America
5 4 3 2 1
CIP data appear at the end of the book

To my daughter Alexa

and my son Peter,

and their families.

Contents

Preface

This book is more than I intended it to be. In 1990, as I was completing a book about the American environmental movement, I recognized that it did too little justice to the contributions and significance of environmental activists around the world. I resolved to follow up with a second book, on international environmentalism.

Then, in the summer of 1990, Maurice Strong, who was visiting in Washington, asked me to join him at breakfast. Strong, a Canadian businessman and diplomat who had been secretary-general of the United Nations Conference on the Human Environment in Stockholm in 1972, had recently been appointed to fill the same role at the 1992 United Nations Conference on Environment and Development in Rio de Janeiro. That meeting soon came to be called the Earth Summit because it would be attended by heads of state and government gathering to discuss the future of the planet and its inhabitants. Strong invited me to join his secretariat as a historian of the process, to be a fly on the wall at the preparatory committee meetings, staff meetings, informal negotiations, his private diplomatic initiatives, and, of course, the summit in Rio itself. I would be given full access to him, his staff, and all conference documents.

To a journalist who had been observing news events from the outside for decades, this offer of an inside view of a major diplomatic conference was an extraordinary opportunity. Even more important, the Earth Summit would bring into sharp focus most of the significant environmental issues facing the global community, as well as assemble representatives of environmental organizations, government officials, scholars, and others who are closely engaged with those issues. Such an arrangement would make my research and interviewing much more compact and manageable. Or so I thought.

I did not, however, wish to join the secretariat. That would have made me a part of the story I was covering, to me a violation of the journalistic ethic. Moreover, I did not want to be in the position of having to submit what I had written to an employer for editing and change. So I requested that I be given the promised access unconditionally. To Strong's great credit, he agreed immediately and kept to his word to the end. I raised money from several foundations (acknowledged elsewhere in these pages) and was able to pursue the research and writing independently.

It became apparent very quickly that if I was going to focus on the Earth Summit process there was no way I could possibly write about environmentalists and environmental issues without also writing about economics and development. Indeed, as I progressed in my research, it became apparent to me that in my years of covering environmental issues I had paid far too little attention to the economic causes of environmental decline and to the environmental causes of economic dysfunction, particularly in the poorer countries of the world. But that was only part of the added burden I was having to take on. It also became evident that if I was going to write intelligibly about environment and development I would have deal with the whole matrix of issues and ideas in which they are enmeshed—science, technology, politics, diplomacy, the institutional and legal framework, justice and equity, and individual beliefs, aspirations, values, and prejudices. As my cartons of documents and transcribed interviews piled higher and my reading list of required books continually expanded, I became dismayingly aware of the truth of John Muir's dictum that everything in the universe is "hitched" to everything else.

I began work on this book after the Berlin Wall was torn down and just before the collapse of the Soviet Union. With the end of the rigid bipolar confrontation of the Cold War and the explosion of freedom and democracy around the world, great new opportunities appeared to be opening for the human community to make a fresh start in the way they lived on this planet, in the way they treated the Earth and each other. Al Gore, a U.S. senator who would soon become vice president, wrote a book proposing "a new central organizing principle for civilization" based on cooperation to preserve the global environment.

At that time, someone sent to my office a stark black-and-white poster with a quotation from Pope Paul VI that proclaimed: "Development is a New Name for Peace." That, I thought to myself, is just about right, provided the word "sustainable" were inserted before "development." I sent away for the 1967 encyclical letter from which the quote was taken, *Populorum Progressio—On the Development of Peoples*, and found that the exact quote was "Development is the new name for peace." The letter went on to state that "excessive economic, social and cultural inequalities among peoples arouse tensions and conflicts and are a danger to peace . . . peace is something that is built up day after day, in the pursuit of an order intended by God, which implies a more perfect form of justice.[1] In 1989, twenty-two years later, Paul's successor in the Holy See, John Paul II, sent out a Christmas message titled "Peace with God the Creator, Peace with All of Creation." In it he wrote: "World Peace is threatened not only by regional conflicts and by injustices between peoples and nations but also by the lack of necessary respect for nature, by the disordered exploitation of her resources, and by the progressive deterioration of the quality of life. The ecological crisis has assumed such proportions as to be everyone's responsibility."[2]

So I had a title for my book: *A New Name for Peace*. To its subtitle, "International Environmentalism, Sustainable Development, and Democracy," I could just as well have added the words "Justice and Equity," but I felt things were getting out of hand.

As it has turned out, this book is a report on the progress, or lack of it, by the human community in its struggle toward that kind of peace—a peace built on mutual care for the planet and responsibility for the well-being of others with whom we share it. To set the context for that examination, I summarize the current threats to the global environment and economy and how they arose. But that is not the purpose of the book. Many other books have already raised those alarms. I also review many of proposals for dealing with those threats, particularly those advanced at the Earth Summit.

My chief objective, however, is to examine what the global community—through its governments and institutions; its political, economic, and social relationships; its science; its intellectual and belief systems; its values; its irrationality; and the everyday acts of its individual members—is doing and not doing about those things that we know are necessary to achieve environmentally sustainable prosperity over the long haul. I have sought to identify doors that have been opened to a postindustrial, postmodern era, an era in which individual and collective security would be defined as preserving the natural world that sustains us and allocating its fruits more equitably among those with whom we share the Earth today and those who will inherit it in the future.

I also, however, have had to devote many pages to describing the obstacles and bottlenecks that are blocking us from crossing the threshholds of those doorways into the beckoning new era. At the end I was unable to conclude that we will be able to overcome those obstacles.

I am a journalist, not an academic. In my early chapters on environmental history, economics, science, and diplomacy, I rely heavily on works by Hayward R. Alker Jr., Edith Weiss Brown, Janet Welsh Brown, Lynton Keith Caldwell, Nazli Choucri, Robert Costanza, Herman Daly, Peter M. Haas, David Lowenthal, John McCormick, John Passmore, David Pearce, Clive Ponting, Gareth Porter, Gwyn Pryns, Caroline Thomas, and Donald Worster, among many other distinguished scholars. The rest of the book is based chiefly, but not entirely, on my own reporting.

In the preface to my earlier book about the American environmental movement, I had also noted that I am a journalist, not a scholar. I was interested to read in several reviews of that book, including favorable reviews, that I had "admitted" or "conceded" that I was not a scholar. Actually, I was not admitting or conceding anything; I was simply letting readers know my credentials. It has been said that journalism is "the first draft of history." A case can be made, in fact, that journalism is closer to the truth than history, if only in a temporal sense.

I found that my training as a reporter with *The New York Times* for thirty-two years was valuable preparation for writing a book that merges as many overlapping issues as this one does. Having covered politics, diplomacy, development, security issues, and diverse cultures as a foreign correspondent in Europe and Asia and having been assigned to write for various periods of time about economic policy, business, trade, labor, science, public health, the White House, and, for my final fourteen years with the paper, the environment, I found the task of synthesizing all of those subjects as they applied to this volume somewhat less daunting than might otherwise have been the case. But only somewhat less.

Chevy Chase, Maryland P.S.
July 1995

Acknowledgments

Generous help from many quarters made his book possible, and I can list only a fraction of those who assisted me on the way.

I wish, first of all to thank Maurice Strong, secretary-general of the United Nations Conference on Environment and Development, who opened the inner doors of the Earth Summit process to me, and who, despite his insanely crowded schedule, always had time to share his knowledge and wisdom. Among the other members of the conference secretariat who shared their insights and information with me were Nitin Desai, Jean-Claude Faby, Nicholas Sonntag, Janos Pasztor, Alicia Barcena, and Nay Htun. Long conversations with Pranay Gupte, editor of the *Earth Times,* helped me think through the maze of issues involved in the summit. So, too, did advice from a number of "wise men" informally working with the secretariat, including Peter Thacher, Pierre Marc Johnson, Richard Elliott Benedick, and Jim McNeill.

I especially want to express my gratitude for the companionship and unfailing kindness throughout the long summit process of the late Benjamin Read, head of Eco-Fund. Ben Read was one of the most thoughtful, decent human beings I have encountered over a lifetime and a selfless public servant.

Janet Maughan of the Ford Foundation not only obtained a generous grant for me from Ford but acted as my guide as I sought the funding required to undertake this project. I am also very grateful to Marianne Lais Ginsburg and the German Marshall Fund of the United States, Wade Greene and Rockefeller Family Services. Dr. Patricia Rosenfield and the Carnegie Corporation, and Margaret O'Dell and the Joyce Foundation of Chicago. I am especially grateful to my wife, Alice Shabecoff, who took time from her crowded schedule to read my manuscript and advise and help me in many other ways.

My sincere thanks to Morris (Bud) Ward, head of the Environmental Health Center of the National Safety Council, and his staff, who generously served as the managers of the funds I received from the foundations, as they did for my earlier book.

Maurice Strong also took the trouble to read my manuscript carefully and to offer advice and point out mistakes. So, too, did Janet Maughan, Stephen Viederman of the Jesse Smith Noyes Foundation, and Konrad von Moltke of Dartmouth College. Their contributions have made this a much better and more accurate book. Whatever errors remain are mine alone.

Finally, I would like to acknowledge the valuable and enthusiastic research assistance I received from Karen Morris Vincent and Patricia Walsh.

A New Name for Peace

Chapter 1

The View from Corcovado

And before him shall be gathered all the nations
—Matt: 25:32

Standing serenely atop the steep peak of Corcovado, the figure of Christ the Redeemer stretched out his arms in an embrace of all-encompassing compassion over a sunny, excited, seemingly happy and prosperous Rio de Janeiro.

At the beginning of June 1992, Rio was the temporary capital city of the world, host to the United Nations Conference on Environment and Development, the biggest, most ambitious summit meeting in history. More than one hundred heads of state and government were en route to Brazil to address what until very recently had been a somewhat arcane subject: environmentally sustainable global economic development.

But the dilemma that the world leaders were coming to address—the interlocking global crises of ecological and economic decline—was far from esoteric. Their task was no less than to plan a new way for the human community to live on this planet.

The meeting was taking place at what appeared to be an extraordinarily fluid and propitious moment in the history of nations. The long Cold War between the Soviet Union and the United States had just ended with the sudden and dramatic collapse of the Soviet empire. The wall—the iron curtain—between East and West disappeared, both physically and symbolically. Communism ceased to be regarded as a viable alternative to free-market capitalism. For the first time in centuries, the planet was free of great-power competition for colonial expansion or military ascendancy. The axis of international confrontation was shifting from an East-West nuclear standoff to a less apocalyptic if more unstable North-South discord. A new chapter in the long chronicle of geopolitics was clearly opening, with blank pages ready to be filled in by the governments and peoples of the world. As Thomas Pickering, then U.S. ambassador to the United Nations, observed, "This kind of opportunity comes along once in a thousand years."[1]

At the same time a slow, quiet, but profound intellectual transformation was changing the way people looked at the relationship between human beings and nature.

For centuries, certainly since the Enlightenment, the prevailing worldview, particularly in the West, held that the destiny of humans was to conquer nature, to use the natural world for their own well-being. "The world is made for man, not man for the world," proclaimed Francis Bacon, the father of modern science, some four hundred years ago.[2] Similarly, René Descartes insisted that men were to be the "lords and possessors of nature."[3] Even in the twentieth century, nature continued to be regarded as a tooth-and-claw enemy of *Homo sapiens,* only a few thousand years from lighting fires at the mouth of his caves to ward off fierce predators and other evils of the night. To Sigmund Freud, "the principal task of civilization, its actual *raison de'être,* is to defend us against nature."[4] It is an article of belief built by the Age of Reason atop the edifice of Judeo-Christian tradition that man was ordained by God to subdue the earth and make it fruitful. While humans are required to be good stewards of the land, they are apart from and above nature and exercise dominion over it. Since the industrial revolution, an almost universally accepted corollary to this creed of human ascendancy has been that the natural world is an cornucopia that requires only manipulation by science and technology to provide an inexhaustible source of sustenance and wealth to an ever increasing human population. The very definition of progress came to mean continued economic and technological expansion without regard to any impact on the biosphere. In the modern world, its civilizations founded on the power of machines and the energy sources that drive them, this mechanistic, anthropocentric perspective has dominated the way much of humanity has approached and used the natural world.

But over the past two hundred years or so a very different concept of mankind's place in world has slowly emerged. Humanity is not separate from and above nature but part of and completely dependent on the natural world, in this view. Human welfare is based not on the conquest and manipulation of nature but on preserving and living in harmony with it. This ecological, biocentric ethic has many roots, including the religions of Asia and Africa. In the West it began to penetrate intellectual circles through the writings of naturalists such as Gilbert White of Selborne, the eighteenth-century English cleric whose precise but loving descriptions of the flora and fauna of his native countryside inspired generations of ecologists. The idea that nature was vital to the well-being of humans was taken up by Rousseau, by Wordsworth and the romantic poets, and by the transcendentalists of New England, particularly in the coolly passionate writings of Henry David Thoreau. Charles Darwin's *On the Origin of Species* presented a compelling scientific rationale for the argument that mankind is an integral part of the chain of being—no more, no less. The mystic and naturalist John Muir, the ecologist Aldo Leopold, and James Lovelock, a propounder of the Gaia thesis that the earth is a single living organism, were among those who urged a holistic approach to the natural world, a world that humans wounded at their own peril. Leopold called for a new "land ethic," which, he said, "changes the role of *Homo sapiens* from conqueror of the land-community to plain member and citizen of it."[5]

As the twentieth century progressed and technology created tools of enormous power and potential destructiveness, a growing number of scholars and scientists warned that humans were, in fact, placing themselves in grave jeopardy by overexploitation and devastation of the natural world. Complaining that the "megamachine" of technological society was enslaving rather than freeing those it was created to serve, Lewis Mumford asked: "What is the use of conquering nature if we fall prey to nature in the form of unbridled men."[6] And in the last decades of the century, a rapidly expanding, activist, global environmental movement, recruiting its forces from the swelling ranks of men and women alarmed by the threat pollution posed to their health and saddened by the degradation of their surroundings, thrust its cause into a place of prominence on the political agenda. To the political scientist Lynton Keith Caldwell, the emergence of the international environmental movement represents "an awakening of modern man to a new awareness of the human predicament on earth."[7]

As a new century and a new millennium approach, the ecolate worldview is posing a strong challenge to the materialistic, technocratic, human-centered model that has long guided the practical affairs of individuals, societies, and nations. While some contemporary thinkers, notably the Belgian Nobel laureate chemist Ilya Prigogine and his followers, continue to insist that humans and other living beings are exempt from the laws of thermodynamics,[8] this is increasingly a minority view.

At the same time, the mainstream of the environmental movement has recognized that its cause is not absolute. In the real world, the pristine wilderness and the pastoral garden have largely vanished. There can be no turning back to a primitive innocence. Preserving nature and preventing pollution cannot be accomplished in isolation from serving the material wants of humanity. The needs of five billion people today and the more than ten billion that will soon inhabit the planet will require all that science and technology can provide to keep them fed, housed, clothed, and healthy. So, too, will the tools that only science and technology can provide be required to repair the ravages that the unwise and arrogant use of science and technology have inflicted on the earth and its inhabitants. Many environmentalists now accept that some form of continuing economic growth, if it is accompanied by a more equitable distribution of the wealth it creates, is imperative to correct the injustice that keeps one-fifth of the earth's inhabitants at the edge of survival while a minority live in comfort and even luxury by preempting the lion's share of the world's resources.

The standard-bearers of the old industrial, technological faith—governments, corporations, economists, scientists, engineers—are also having to accommodate their beliefs to the reality of shrinking resources, pollution that affects production efficiency as well as health, and rising threats to the atmosphere and other ecological systems that support life on earth, including their own lives. Governments have long since recognized that environmental dangers do not respect national borders, no matter how well fortified. A growing number of businesspeople and economists now agree that, to sustain economic growth over the long run, resources will have to be better husbanded and the real costs of pollution will have to be accounted for. Scientists and engineers are acknowledging that their work can have unintended destructive effects on the natural world, effects that can have profound consequences for the future of both nature and humans.

In their book *Only One Earth,* written in 1972 for the UN Conference on the Human Environment in Stockholm, the economist Barbara Ward and the microbiologist René Dubos wrote: "The two worlds of man—the biosphere of his inheritance, the technosphere of his creation—are out of balance, indeed potentially in deep conflict. And man is in the middle. This is the hinge of history at which we stand, the door of the future opening to a crisis more sudden, more global, more inescapable and more bewildering than ever encountered by the human species and one which will take shape in the life span of children who are already born."[9]

Such an assessment of the predicament facing the human community, regarded by many as alarmist in 1972, was widely acknowledged as all too accurate twenty years later. The once antithetical biocentric and mechanistic worldviews—historian Donald Worster terms them "arcadian" and "imperialist" approaches to nature[10]—were, by the time of the June 1992 Earth Summit, slowly converging. The environment and economic growth need not be in conflict. In fact, by the time of the summit it was understood that, without protection of ecological systems, global economic decline was inevitable. Conversely, without economic progress, elimination of poverty, satisfaction of the material wants of people of the developing countries, and extension of human rights, efforts to protect nature and the earth's life-support systems were doomed to failure. A concept called sustainable development, which had been slowly entering the language of geopolitics over two decades, could embrace both views of how the world actually works.

. . .

Given the startling new prospects opened to diplomacy by the end of Cold War and by the intellectual breakthrough on the human relationship to nature, hopes were high for the Rio Conference. Just possibly, it would lay the foundation for a new system of collective security, replacing the armed bipolarity of the previous half century with a system of global cooperation to preserve the environment while building an equitably shared global prosperity. Just possibly, geopolitics could be restructured around what then-Senator Al Gore, in his book *Earth in the Balance,* described as a new "central organizing principle for civilization."[11] Just possibly, the endless struggle for world peace would be based, in the future, not on strength of arms, military alliances, or the exercise of economic power but on loyalty to and a shared concern for the fate of what Sir Shridath Ramphal, former secretary-general of the British Commonwealth, called his book, *Our Country, the Planet.* "The war for human survival," he wrote, "is unlike other wars. . . . It is not a war of man against man, nation against nation, but rather a war of humanity against unsustainable living."[12]

As thousands of delegates, journalists, environmentalists, and other lobbyists streamed into Rio, portents seemed favorable. The city itself was a sparking, hospitable, orderly tropical paradise—a city of blue lagoons, glittering white beaches, palm trees, mosaic sidewalks, lithe young beauties in string bikinis; of the Copacabana, Ipanema, Sugarloaf, and smiling, friendly *cariocas*. The violent crime that had been disrupting the city was nowhere to be seen. The *peixote*, tens of thousands of homeless children who slept in the streets, begging, scavenging and stealing to stay alive, also had seemingly vanished. There was little trace of the extreme poverty suffered by much of the city's population. A democratic, apparently stable government ruled a country once dominated by military juntas. The visitors were arriving in what looked very much like the old Rio of Hollywood—the carefree Rio of Carmen Miranda, of Astaire and Rogers, Hope and Crosby.

But in reality, the dark side of Rio had been painted over for the conference with a thin, temporary coat of whitewash. Crime and violence were held at bay only by extreme security measures, which included heavily armed troops lining the streets and a greatly augmented police force, with contingents assigned to every major hotel. Many of the street children reportedly had been rounded up by police and forcibly detained in a neighboring city across the bay from Rio for the duration of the meeting. The poverty was concentrated in the *favelas*, the shantytowns clinging precariously to the cliffs high above the city, far from the conference, the hotels, the elegant restaurants. These crowded slums are dangerous concentrations of filth, hunger, drug addiction, and despair—and also, it must be said, of much laughter and music. But from a safe distance they could be seen as pastel clusters of picturesque cottages that only made the setting more exotic. And even as the world leaders descended on the city, Brazil's president, Fernando Collor de Mellor, was on the verge of toppling amid charges of widespread corruption.

Just as Rio was something other than it seemed on the eve of the Earth

Summit, so, too, was the international political climate not nearly as promising as might appear at first glance. The breakup of the Soviet Union unleashed ancient ethnic, religious, and nationalistic antagonisms, which led to violent and bloody conflicts in the Balkans and the other parts of the former Soviet empire. The economies of Russia and many of her former satellites were *in extremis.* The United States and its allies had to fight a war to deny Iraq its conquest of Kuwait and that country's rich oil fields. Despite the almost miraculous agreement between Israel and the Palestinians to acknowledge each other's existence and bargain a peace agreement, there seemed to be no end to the violence and hatred in the Middle East. Indeed, the rise of militant Islamic fundamentalism seemed to augur even more instability and bloodshed in that deeply troubled region. Although apartheid had finally ended and there was high hope for the multiracial government in South Africa, much of the African continent was in prolonged economic and environmental decline. The Horn of Africa, Rwanda, Burundi, Liberia, and some other regions were beset by famine and insanely brutal factional strife. The Indian subcontinent and Sri Lanka also were roiled by religious and ethnic hatreds. Human rights abuses in China slowed that nation's reentry into the community of nations. Much of Latin America continued to be beset by poverty and the maldistribution of wealth, with added problems caused by drug production, trafficking, and addiction in a number of those countries.

Also damaging to prospects for the Earth Summit, was the prolonged if sporadic recession that was afflicting the economies of United States, Western Europe, and even Japan. Beset by pressing, seemingly insoluble domestic problems, the leaders of the industrialized democracies were in no mood for bold new international programs as they arrived in Brazil, particularly if those programs were to involve the expenditure of any substantial amounts of money. Moreover, virtually all of the heads of government in those countries had proved to be weak leaders, and corruption was rotting some of these governments from within.

Finally, delegations were coming to Rio with widely differing agendas. Most of the rich industrialized countries were concerned chiefly with international and global environmental issues, particularly the potential environmental dangers that would arise from industrial development in the poorer countries. The developing nations, on the other hand, wanted the conference to endorse substantial new economic assistance: increased bilateral and multilateral aid, better terms of trade, debt relief, access to technology. Several countries, including the U.S. administration of President George Bush, clearly were hoping that the conference would not produce any major changes for existing international relationships and responsibilities. China viewed the conference as an opportunity to emerge from its self-imposed diplomatic isolation.

Thus, the Rio Conference was like the city itself—a promising, even dazzling, surface covered a reality that was riddled with strife and hidden dangers.

Failure, however, was not an acceptable option for the nations of the world. The stakes were much too high. Threats to the future of human welfare had reached a magnitude that could not be ignored, even by the most insular and inward-looking governments. Shortly after the conference ended, a "Warning to Humanity" was issued by 1,575 prominent scientists from sixty-nine countries, including a majority of living Nobel laureates. The "warning" stated, in part:

> Human beings and the natural world are on a collision course. Human activities inflict harsh and often irreversible damage on the environment and on critical resources. If not checked, many of our current practices put at serious risk the future that we would wish for human society and the plant and animal kingdoms and may so alter the living world that it will be unable to sustain life in the manner that we know. Fundamental changes are urgent if we are to avoid the collision our present course will bring about.[13]

The threats that are placing the natural world and its human and nonhuman inhabitants at risk and that led the UN General Assembly to call the Rio conference are by now familiar to most readers, and there is no need here to recount them in detail.

Many of those problems stem directly or indirectly from the growth of human population, now over five billion and heading to at least ten billion. Some long-term projections postulate a global population of as much as twenty billion by 2150 without strong measures to slow fertility rates.[14]

In his book *Preparing for the Twenty-First Century*, historian Paul Kennedy reported that "the over all consensus—with the exception of a few revisionists—is that the projected growth in the world's population cannot be sustained *with current patterns and levels of consumption* [italics in original]."[15] Dr. Nafis Sadik, executive director of the UN Population Fund noted before Rio that the debate over sustainable development and protecting the global environment "will be meaningless if population is ushered to the sidelines. Population issues are central to the search for sustainability."[16]

Human numbers, combined with our technology and economic activity have reached the dimensions and power equal to elemental forces of nature. We have already begun to put excessive strain on the systems that make this planet habitable—the atmosphere, the land, the oceans and fresh waters, the plants, animals, and microorganisms. This stress is beginning to exceed the absorption capacity of these systems and limits their ability to meet our needs over the long run. We are already experiencing surprises from the effects of massive human intervention in the biosphere. Global warming, the degradation of the protective ozone layer, the decline of biological diversity, the loss of soil and forests, the contamination of our fresh water supplies and even the great oceans, crashing fisheries, the flood of toxic and radioactive substances entering our environment and our bodies and threatening our physical and reproductive health—all signal that humans are making excessive demands on the global habitation that sustains us.

These assaults on the environment pose dangers for our health, our ability to sustain ourselves over the long run, the quality of our lives, and indeed, the future of life on earth. Many, like those who signed the scientists' "Warning to Humanity" quoted above, believe that time to address these threats to the physical world is running out. Shortly before the Earth Summit, U.S. Senator Tim Wirth asserted that "global environmental decline—or the preservation of God's creation—is the great moral and ethical question of our time. Survival is at stake—the answer and the obligation is to begin to turn things around. This enormous responsibility overwhelms all others. We have no choice but to live up to it, to its challenge, to its opportunity. The environment is a superpower."[17]

But environmental decline stems largely from our industrial and agricultural activities—the things we do to support ourselves; to put food and drink on the table; to house, clothe, and warm ourselves; to travel; to communicate; and to entertain ourselves. We cannot simply stop. We cannot step out of the speeding jet aircraft. We cannot stop producing food, building houses, burning coal and oil, making automobiles, and using synthetic substances. We cannot stop our efforts to reduce poverty and remove the great gulf that separates the rich and poor people of this world. Nearly a billion people today, one of every five human beings, continue to live at the margins of survival, with barely enough food to keep alive, without clean drinking water or adequate shelter. Global security, not to mention simple justice, requires that the international community take the economic measures necessary to lift this mass of humanity out of extreme want and give their lives some portion of dignity. We cannot—or should not—tell people in the developing countries who aspire to a higher standard of living to refrain from the patterns of consumption and waste that we in the rich nations continue to follow. Even in the wealthy countries—in North America, Western Europe, Japan—economic growth is needed to provide jobs and the necessities and comforts for growing populations.

In 1987 the World Commission on Environment and Development issued a report titled *Our Common Future* but usually called the Brundtland report after its chairwoman, previously and later the prime minister of Norway, Gro Harlem Brundtland. It warned that the planet was beset by the "interlocking crises" of environmental and economic decline. Explosive population and economic growth were putting unsustainable pressures on the natural environment, which in turn was piling up increasingly difficult obstacles in the path of future economic and social development. "Ecology and economy are becoming ever more interwoven—locally, regionally and globally—into a seamless net of cause and effect," the report cautioned.[18]

. . .

This is not, however, a book about problems. It is about how people and governments, educated by science and prodded by an emerging global environmental movement, have gradually come to recognize the threat to the natural world and to themselves as a result of human activity; how our institutions, laws, politics, diplomacy, economics, science, and the broad society have responded—or have not responded—to this slowly dawning awareness; where we stand in the last years of the twentieth century; and the prospects for altering the present collision course and meeting the challenge of creating an ecologically rational, prosperous, and just global economy and a new system of collective security among nations.

As they descended on Rio, the world leaders faced a large number of extraordinarily difficult questions. How can the imperatives of economic growth be reconciled with the increasingly urgent task of protecting planetary life-support systems? How can development be conducted so that it is sustainable into future generations? How, given the great gulf between the needs and demands of the rich and poor countries, can the global community refashion its economic relationships? How, if development were broadly achieved, could the world sustain for ten billion people the material consumption and waste production that accompanies the life-styles in the handful of rich industrialized nations? Can there be a new kind of growth that does not degrade and diminish the physical world? These were the puzzles that global leaders were being asked to solve as they descended on Rio.

Such questions posed problems that international policymakers have not really had to address in the past in any serious, sustained way. Sovereign nations rarely have had to deal with the impact of purely domestic activities on other countries, nor indeed, on the planet as a whole. Nor have they had to plan economic and industrial policy to account for their effect on the environment fifty or one hundred years from now and their effect on future generations. They have not been required to make critical, long-range decisions on the basis of so much scientific uncertainty.

They had had little precedent to guide them. Green diplomacy was still a slender chapter in the history of nations. Global environmental politics was still in its infancy. The voice of democracy was rising but still uncertain in much of the world.

The Earth Summit was a rare moment of opportunity for the human community to change direction, to correct some of the structural flaws that create human misery and that were beginning to threaten the very systems that support life on earth.

But the world leaders en route to Rio were carrying an overwhelming burden of intellectual, cultural, and institutional baggage that would weigh them down as they grappled with the sweeping agenda of the conference. They would have to address the economic causes of environmental decline within the context of a two-hundred-year-old economic system that simply does not account for such problems and does not contain the tools for solving them. They would need to construct their decisions on a base of sound science—but science was becoming increasingly commercialized, politicized, and adversarial. The traditions of diplomacy under which they would face the complex array of issues stressed the exercise of power and competition for national advantage rather than the cooperation and sharing needed to deal with the threats to the planet. Despite flashes of promise, the UN had failed to evolve into the center of international governance essential for global cooperation. The widest possible public consent would be required to support the decisions of world leaders; but geopolitics had traditionally been conducted behind closed doors, and its practitioners had little experience in working with environmental groups and other democratic institutions that sought to make their voices heard at the Summit.

Around the world, many people looked with hope toward Rio as a portal to a new, post–Cold War, postindustrial era of collective security based on cooperation to safeguard the planet and to keep it habitable and hospitable for future generations. But the way through that portal was obstructed by a labyrinth of human history, systems, institutions, economic and political arrangements, social and legal structures, and flawed and hostile relations among nations.

What follows is an effort to draw a rough map of that labyrinth, to examine the obstacles and roadblocks strewn along the way to a more secure and livable planet—and the possible paths around them.

Chapter 2

The End of Innocence

. . . I led men on the road
Of dark and riddling knowledge; and I purged
The glancing eye of fire, dim before,
And made its meaning plain. These are my works.
—Aeschylus, *Prometheus Bound*

The annals of international cooperation to preserve and protect the environment are recent and brief. Indeed, the very notion that the natural world must be defended against human assault, even at the national and local level, is a relatively new concept. A century ago, any suggestion that human activity could cause permanent destruction on a global scale and threaten the future of life on earth would have seemed startling and outrageous. Physical evidence of consequential environmental degradation was scant, scattered, and the concern of relatively few scientists and amateur naturalists. Little need for international environmental action was perceptible. And while the relationship of humankind and nature has been pondered by scholars, scientists, theologians, and ordinary people for millennia, there was no intellectual tradition that could admit of men harming themselves by harming nature. In a geological and evolutionary time frame, *Homo sapiens* has only recently emerged from its caves, where it huddled around a meager fire, struggling to survive in a world of fierce, hostile, overwhelmingly dominant nature.

. . .

To the ancients, nature was created and controlled by (and largely composed of) the gods. The earliest creation myths have the gods molding an ordered nature out of chaos, usually, but not always, for the benefit of humans. These tales arose from the experience of the first civilizations as they groped toward survival and stability in an untamed, threatening landscape. Thus, the creation myths of the Sumerians and Babylonians reflected their struggle to grow crops and build cities on the watery marshes of the Fertile Crescent.

> When the mother of the gods had been created,
> When the earth was created and fashioned,
> And the destinies of the sky and the earth fixed,
> When ditches and canals had received their proper course
> And the banks of the Tigris and Euphrates had been established.[1]

Having fashioned nature, only the gods could assure its beneficence, and only they could damage and destroy it. The ability of humans to transform the natural world depended on their relationship to the gods, or to God. If the gods were pleased, a generous nature would nurture and sustain them. If they created evil, nature would be turned against them. The flood legends of the Sumerians and the Old Testament were moral parables of broad ecological change wrought by a supernatural agent in response to human behavior.[2]

When the Titan Prometheus, taking pity on the weak, shivering race of humans, stole knowledge and fire for them from Olympus, he was cruelly punished by Zeus, who chained him to a pillar in the Caucasus Mountains, where each day his liver was torn by an eagle from his immortal body. To the ancient Hellenes the Prometheus legend held a simple lesson: offenses against the gods are punished, even when the offense serves humanity. Not until very recently has it begun to dawn on us that the offense against the gods for which Prometheus was punished was not the theft but the gift itself. The "dark and riddling knowledge," the titanic energy and technology presented to humans by Prometheus, gave them power—exceeding the power of Zeus himself—to use, change, degrade, and in our time, destroy the natural world.

As humans began to grapple with who they were and where they came from, one of the early teleological conclusions was that all of their universe, including themselves, was the work of some supreme force or creator. Humans were not, therefore, masters of the natural world but humble parts of it. To the historian Clarence Glacken the idea of an earth of nature and man designed in toto by one or more creator/artificers was "one of the great attempts in Western civilization, before the theory of evolution and modern ecological theories emerging from it, to create a holistic concept of nature."[3]

While this belief was to have a lasting impact on Western thought, other concepts of the relationship between man and nature gradually emerged, Glacken noted. One was that the environment, however created, could shape and mold human beings and their societies. Climate, weather, and variations in flora and fauna helped to determine the nature of men and nations, in this view.

Another conclusion reached by humans as they grappled with the intellectual puzzle of their place in the world was that man himself was an agent of change in and modification of the natural world.[4] But for millennia there was no question that those changes were ordained and benificent.

In Genesis, Adam is presented the right and duty to "be fruitful, and multiply, and replenish the earth, and subdue it: and have dominion over the fish of the sea and over the fowl of the air, and over every living thing that moveth upon the earth."[5] Humans are thereby given the authority to act as God's agents in using and modifying the environment. Of course, the Bible also admonishes humankind to "hurt not the earth, neither the sea, nor the trees."[6] But for most of history, human intervention in the natural world was regarded either as benign and beneficial or, at worst, neutral.

The Judeo-Christian tradition that placed man apart from and above nature was reinforced by the scientific materialism of the Enlightenment, which defined human progress by the knowledge and conquest of the natural world. "Few enjoyments," wrote the philosopher David Hume in the mid-eighteenth century, "are given to us from the open and liberal hand of nature; but by art, labour, and industry, we can extract them in good abundance."[7] The mechanistic, utilitarian theories of Enlightenment thinkers such as Bacon, Descartes, Hume, and John Locke led to the "desacralization of nature," according to the social critic Jeremy Rifkin.[8] Rifkin contends, however, that the alienation of humans from nature began, in Europe at least, somewhat earlier, with the dismantling of the commons. "The enclosure and commodification of the global commons changed humanity's long-standing relationship to the land. . . . The land was no longer something people belonged to, but rather a commodity people possessed."[9]

There were, of course, repeated challenges to the faith that human conquest of nature was divinely favored, as well as scientifically and economically rational, and that it served human welfare in a well-ordered, moral universe. Jean Jacques Rousseau, who wrote at the same time as Hume, argued that "the earth left to its own natural fertility and covered with immense woods, that no hatchet ever disfigured, offers at every step food and shelter to every species of animals."[10] It was "iron and corn"—that is, industry and agriculture—"which first civilized man and ruined humanity," Rousseau asserted.[11] An emerging romanticism was making nature the object of veneration and the source of wisdom and virtue, as illustrated by these lines penned by William Wordsworth:

> One impulse from a vernal wood
> May teach you more of man,
> Of moral evil and of good,
> Then all the sages can.[12]

But to most people, struggling to survive and prosper, raw nature was still an enemy to be conquered or, at best, a force impervious to the feeble strength of humanity. To their eyes a vernal wood was lumber for houses and firewood, trees to be cleared to make way for crops and cattle. In the eighteenth century, industry, agriculture, and other human works were still, in the main, leaving only superficial, transient scars on the body of the natural world. A population that was well short of a billion people, living in a world with still vast reserves of open space, using tools of wood and iron that could only scratch lightly at the surface of the earth, and depending chiefly on the muscles of men, women, and animals, plus fire and water, as sources of energy, was living more or less on the boundaries of what nature could continue to supply indefinitely.

To be sure, there were significant man-made changes in the natural landscape in preindustrial, even prehistorical times. As historian Clive Ponting has pointed out, several great ancient civilizations, including those of Mesopotamia, the Indus Valley, and Mesoamerica, collapsed when excessive populations and ultimately destructive agricultural practices placed unsustainable pressure on their basic ecological systems.[13] Great swaths of Europe and Asia were deforested. Of England's original forest cover, for example, only 20 percent remained by the time the Domesday Book was assembled in 1086 (by the beginning of the twentieth century only 4 percent was left).[14] Some 6,500 years ago, the city-states of Lagash and Umma went to war over what already had become scarce water supplies in the region we now call the Middle East.[15] The eradication of much of the wildlife of the heavily settled areas of the Old World and decline of once fertile areas of the Middle East into deserts are but two older examples of human ability to wound nature.

The impact on the lives and economies of people, however, was local or regional and made very little impression on the broad human consciousness. Whatever rudimentary steps were taken by governments to preserve natural resources, such as protection of fishing waters or restrictions on the taking of wildlife, were done out of economic interest or to protect the privileges of the noble and landed classes. In 1337 and again in 1363 the British Parliament, in order to preserve disappearing fur-bearing animals and to keep furs as a symbol of status, passed largely ineffective laws to limit the wearing of desired furs to the nobility and clergy.[16]

But the general Western view, held until very recently, was expressed by the great scientist Wilhelm von Humboldt: "Nature goes forward in her never-ending course, and cares nothing for the race of men that is ever passing before her."[17]

At the beginning of his monumental *Civilization and Capitalism* the historian Fernand Braudel speculated that in all ages there is a "limit" that places restraints on human attainment, a "borderline" separating the possible and the impossible. In preindustrial economies, he wrote,

> the borderline was imposed by inadequate food supplies, a population that was too big or too small for its resources, low productivity of labour, and the as yet slow progress in controlling nature. Between the fifteenth and eighteenth century, these constraints

hardly changed at all. And men did not even explore the limits of what was possible. . . . In the end, the only real change, innovation and revolution along the borderline between the possible and the impossible came with the nineteenth century and changed the face of the world.[18]

That change was the result of an industrial and agricultural revolution based on a continuing series of astonishing scientific and technological innovations and responding to expanding markets at the continental and global level. The explosion of knowledge and the Promethean new sources of energy—and the powerful new machines and tools they produced—in many ways bettered the human condition over much, but not all, of the world. They eased the burden of labor, opened a cornucopia of food and other material goods, reduced the terror of disease and pain, prolonged human life, and shrank the size of the globe with high-speed transportation and instantaneous communication.

In time the industrial revolution also created profound transformations in the physical world. Advances in food production, medicine, and hygiene opened the way to the vertical leap in population over the past two hundred years, particularly in this century. The new machines and the energy sources that fueled them enabled forests to be cut down with accelerating speed, opened huge new areas of the land to cultivation, exposed the depths of the ocean to the nets of fishing fleets, permitted mines to be sunk deeper into the earth and mountains to be leveled for the mineral riches they contained, and allowed humans to cover ever more of the natural landscape with their cities and suburbs, highways and railways, industrial complexes, dams, canals, bridges, power lines, and communications towers. The synthetic chemicals, wastes and effluents of an increasingly productive society, many of them hazardous and even deadly, seeped into the ground, the drinking water, and the flesh of animals and fish, into mothers' milk and into the living cells of almost every human on earth.

War and conquest also were industrialized, greatly increasing the efficiency with which men could kill each other—and civilian populations—and facilitating the rapid spread of colonialism in Africa and Asia. It also permitted, in one scholarly view, "the competitive, expansionary, modern practice of seeking dominance over a femininely gendered 'nature' (including 'inferior' peoples and resource bases anywhere around the globe)."[19] The gradual eradication of the native landscape in Africa, Latin America, North America, and parts of Asia progressively eroded the hunt-and-gather or slash-and burn economies of indigenous people in those areas and, tragically often, completely eliminated those societies. But as Ponting pointed out, "Europe's success in breaking free from the long struggle to survive lay in its changing relationship with the rest of the world and, in particular, its ability to control an increased share of the world's resources."[20] The demand for standardized parts for weapons was also one of the incentives for the development of the modern mass-production assembly line.

This ever accelerating activity, spurred by economic systems that encouraged high levels of consumption and continual economic expansion, began to take an increasingly ominous toll on the air, water, soil, gases, plants, animals, and microorganisms that support life on earth. In our time it is altering the very chemistry of the planet—significantly shifting, for example, the natural carbon, sulfur, and nitrogen cycles. In their essay "The Great Transformation," Robert W. Kates, B. L. Turner II, and William C. Clark note that "whereas humankind once acted primarily upon the visible 'faces' or 'states' of the earth, such as forest cover, we are now altering the fundamental flows of chemicals and energy on the only inhabited planet we know."[21] While fundamental environmental change in the past was caused almost exclusively by natural forces, in recent times, "humanity has become such a force that it is difficult to characterize many changes or transformations of the biosphere independent of it."[22]

. . .

The quantum and still accelerating expansion of human power to change the natural world has been accompanied over the past two centuries by transformations in the understanding of the relationship between humans and nature, albeit a significantly slower transformation and one that is still dangerously far from complete.

Perhaps the greatest stimulus to the intellectual dialogue on the relationship of humans and nature in the nineteenth century was the publication of Charles Darwin's *On the Origin of Species* in 1859. Darwin's painstakingly researched, modestly stated theory of natural selection made it inescapably clear that *Homo sapiens* was totally enmeshed in the web of nature, not apart from or above it. His demonstration that the response of all organisms to their surrounding environment determined their survival and evolution provided the foundation for a new scientific discipline based on the now perceived "biological unity of nature"—ecology.[23] The word itself was coined in 1866, just seven years after Darwin's great work appeared, by the German Darwinian Ernst Haeckel, from the Greek word for household, *oikos*.[24]

On the Origin of Species placed man and nature in the same social context: humans, plants, and animals are all members of the same community of life. To some the chief lesson of natural selection was the imperative of cooperation with nature to protect the ecological habitat—the household—that humans shared with other organisms. Others drew the contrary conclusion that survival of the fittest justified, indeed required, the ruthless exploitation of nature for human needs.

Darwin's descriptions of the brutal struggle for existence as the fate of all life turned many in the Victorian era away from the romantic ideal to a more pessimistic view of nature.[25] But Darwin himself, while occasionally ambiguous about the moral implications of his work, reached a rather optimistic conclusion at the end of *On the Origin of Species:*

> As all the living forms of life are the lineal descendants of those which lived long before the Cambrian epoch, we may feel certain that the ordinary succession by generations has never once been broken, and that no cataclysm has desolated the whole world. Hence we may look with some confidence to a secure future of great length. And as natural selection works solely by and for the good of each being, all corporeal and mental endowments will tend to progress toward perfection.[26]

The depredations of the industrial revolution on the natural world also began to be reflected in evolving political and economic philosophies. Both Karl Marx and Friedrich Engels, for example, equated the exploitation of nature with the exploitation of workers. Thus, Engels wrote: "The earth is the first condition of our existence. To make it an object of trade was the last step toward making human beings an object of trade."[27] But as Richard Barnet has noted, socialism, like capitalism, cannot accept the implications of a Malthusian world of scarcity. Both systems are instead built on a "myth of abundance" that is indifferent to the toll that the creation of that abundance takes on the environment.[28]

. . .

The industrial revolution came to the United States somewhat later than to Darwin's England, but it transformed the pristine North American landscape with breathtaking speed. For a long time most Americans remained indifferent to the leveling of their forests, the disappearance of their wildlife, the fouling of their lakes and rivers, their reeking industrial areas, and the grime and noise of their cities. These were regarded as a small price to pay for the progress and prosperity of their young country. But there were a few who realized that the price was steep and the losses irrecoverable.

One of the earliest and most eloquent voices to be raised in mourning for the loss of nature in, America or elsewhere, was the New England transcendentalist, Henry David Thoreau. To Thoreau, born in 1817, the destruction of the wilderness and the retreat of pastoralism before advancing industrial capitalism were grievous wounds on the soul and spirit of men and women. Only in "wildness" could the world be preserved, he warned. He perceived the world as a living, integrated entity that men defaced at their own peril. "The earth I tread on," he wrote, "is not a dead inert mass; it is a body, has a spirit, is organic and fluid to the influence of its spirit." He lived for two years as a hermit in a cabin on Walden Pond, he explained, because "I wished to live deliberately, to front only the essential facts of life." He was among the first to claim that the natural world—animals, birds, trees, even weeds—had a right to exist for their own sake, not for their utility to humans.[29]

Thoreau can be regarded as the forefather of American environmentalism. His passionate insistence on the spiritual dependence of humans on nature and even some of his prescribed tactics are part of the basic primer of values and techniques of today's environmental movement. But it was another American Yankee, George Perkins Marsh of Vermont, who made the first systematic and influential assessment of the serious and permanent damage that human activity can inflict on the natural world.

Marsh, a lawyer, teacher, farmer, businessman, newspaper editor, linguist, scientist, member of Congress for the Whig Party, and for many years U.S. ambassador to Turkey and then to Italy, observed nature from neither a romantic, transcendentalist perspective nor from a narrow, utilitarian one. He was a careful, deductive observer, and from his observations of the degraded landscapes of southern Europe and the Middle East and from his deep scholarship he produced, in 1864, his great work: *Man and Nature: or, Physical Geography as Modified by Human Action.* In it he described how nature is basically stable, but cumulative human activity such as forestry, irrigation, cultivation, and drainage could cause major, irreversible damage, including the loss of topsoil and watersheds and of plant and animal species, salinization of the land, and even a change of climate in local and regional areas.

"But man is everywhere a disturbing agent," he warned. "Wherever he plants his foot, the harmonies of nature are turned to discord."[30] Marsh asserted, adding that by so doing he inflicted deep wounds on his own interests. Noting, for example, that insect pests proliferate when people kill the birds that eat them, man is "not only depriving his groves and his fields of their fairest ornament, but he is waging a treacherous warfare on his natural allies."[31]

The central lesson of Marsh's work, said David Lowenthal, in his introduction to the centennial edition of *Man and Nature,* is that "man must learn to understand his environment and how he affects it. For his own sake, not for nature's alone, man must restore it and maintain it for as long as he tenants the earth."[32] Marsh himself asked: "Could this old world, which man has overthrown, be rebuilded . . . ?" His answer was that such a rebuilding "must await great political and moral revolutions in the governments and peoples."[33]

Man and Nature was something of a sensation in Europe as well as in the United States. The idea that human activity endangers human welfare by destroying nature was new and disturbing. But its lesson would go essentially unheeded for many decades. The "political and moral revolutions" that Marsh so presciently saw would be long in developing are yet far from achieving their objective. People across the world, preoccupied by the burdens of day-to-day existence, were slow to recognize the danger. Their governments were even more sluggish and oblivious. In part that inaction may have reflected what the historian Theodore Roszak describes as "the helplessness many people of the period felt before the vast, inexorable forces that the industrial revolution had released in the world."[34]

. . .

Still, the roots of modern environmentalism are embedded deeply in the soil of the nineteenth century.

The recognition of waste and of dwindling resources led to the creation of scientific forestry in Germany and the establishment of nature "conservancies" in British India. The world's first private environmental group, the Commons, Open Spaces and Footpaths Preservation Society, was founded in 1865 to campaign for land preservation, particularly urban commons, by English men and women who were revolted by the increasing squalor of urban industrial life.[35] In 1880, English citizens formed the Fog and Smoke Committee to try to combat the air pollution suffocating urban areas.[36] Concern about wildlife, rapidly disappearing in the face of hunting and economic development, spawned many private conservation groups in Britain, Australia, New Zealand, and colonial Africa.

Anger and sadness over the ruthless extermination of wildlife and the exploitation of forests and other public resources also were the earliest stimulus to private environmental initiatives in the United States. In 1887, George Bird Grinnell, who would later found the first Audubon Society, joined with several wealthy sportsmen, including a young politician named Theodore Roosevelt, to form the Boone and Crockett Club, which sought to end the wasteful slaughter of big-game animals, including the rapidly vanishing bison.

Modern environmentalism in the United States emerged at the end of the nineteenth century in the form of two often conflicting philosophies. One was conservation, championed by Gifford Pinchot, the first chief of the U.S. Forest Service, and by his boss, President Theodore Roosevelt. Conservation insisted on a scientific, utilitarian approach to managing natural resources in the public domain. Forests should be exploited for the economic benefits they provide but in ways that could be sustained into future generations. A centerpiece of Roosevelt's progressive approach to governance, conservation embodied his administration's democratic ideals: public resources should serve all the people of the nation, not just, as had been all too often the case, the handful of robber barons who used and brutally degraded them for the purpose of accumulating private wealth. In some important ways the early conservation movement anticipated the emergence of the "sustainable development" ethic a century later.

The other major strand of the emerging environmental impulse in America at the end of the nineteenth century was the preservation movement. Arising out of the transcendental and romantic traditions and propounded most passionately by John Muir, the bearded naturalist and mystic, preservationism sought to protect wilderness and wild nature not only because they were essential for the health of the human soul but because they had an intrinsic right to exist for their own sake. The preservation movement adopted a broad biocentric—as opposed to an anthropocentric—perspective and gave rise to many of the amateur environmental groups formed in the first half of the twentieth century. John Muir himself was a co-founder of the Sierra Club, one of the earliest of the national environmental groups that rallied to the support of a beleaguered nature. For many, preservationism became a quasi-religious cult of the land.

Governments in the nineteenth century were slow to react to the rising toll of industrialism and the increasingly voracious demands of the global market. Responding to the devastation industrial emissions were inflicting on the cities and countryside, the British Parliament passed the Alkali Act of 1863 and, by so doing, acknowledged government responsibility for acting to reduce air pollution. In the 1860s, 1870s, and 1880s, Parliament also enacted a series of laws offering a measure of protection to birds and animals that were being slaughtered for their feathers and furs.[37]

In the United States, the federal government began building the national park system, starting with Yellowstone in 1872, and the national forest system. Many cities began adopting antipollution, sanitation, and public health ordinances. Teddy Roosevelt's administration was the great exception to the general ignorance and indifference to environmental problems demonstrated by most national governments until recent times. In addition to making protection and democratization of public lands and resources a centerpiece of his administration, he also called the governors of the states to the White House in 1908 for a conference on conservation. That conference, which made a major conceptual breakthrough by establishing the protection of human health from the dangers of pollution, has been described as the beginning of a true national conservation movement. But the federal government withdrew as a significant actor on the environmental stage as soon as William Howard Taft succeeded Roosevelt as president.[38]

. . .

If governments and private citizens moved slowly to address domestic environmental ills, the pace of international cooperation, bilateral or multilateral, to deal with wildlife preservation, transboundary pollution, and protection of the global commons was glacial. In 1872 the Swiss government proposed creation of an international commission to protect the birds that migrated across national borders in Europe, but other sovereign European states had no interest and the idea died. Not until 1979 did the Europeans agree on a treaty to conserve wildlife and natural habitat on their continent.[39]

As the legal scholar Edith Weiss Brown has noted, the few international environmental agreements that existed in the nineteenth century and early part of the twentieth century "were based on unrestrained national sovereignty over natural resources and focused primarily on boundary waters, navigation and fishing rights along shared waterways." The Europeans agreed in 1900, for example, on the London Convention for the Protection of Wild Animals, Birds and Fish in Africa, which, of course, posed no threat of infringement on their sovereign rights to exploit their own resources. Brown also pointed out that the international accords did not address pollution or other ecological problems. "The dramatic exception" was the 1909 Boundary Waters Treaty between the United States and the United Kingdom, which required that water not be polluted and health and property not be damaged on either side of the border between the United States and Canada.[40]

One more early attempt to address environmental problems on a multinational front was the North American Conservation Congress called by President Theodore Roosevelt and attended by representatives of the United States, Canada, and Mexico in Washington, D.C., in February 1909. "The most notable outcome of the congress," according to environmental historian John McCormick, "was its agreement that conservation was a problem broader than the boundaries of one nation." Acting on the recommendation of the congress, Roosevelt invited fifty-eight nations to attend a "world conservation Congress in the Hague, and nearly half the invitees accepted. But Roosevelt left office before it convened and it was canceled by President Taft.[41] With it went glimmering an early opportunity to establish the framework of an international regime to protect the global commons.

For most of the first half of the twentieth century, disrupted by two murderous world wars and a long period of global economic decline, public attention rarely focused on environmental matters, and public policy for the most part ignored them.

There were some exceptions. In the United States the appalling erosion and drought that carried away millions of tons of topsoil from the Dust Bowl in the 1930s called alarmed attention to the ecological effects of human economic activity. The administration of President Franklin D. Roosevelt responded with a series of conservation initiatives, including the Soil Conservation Service and the Civilian Conservation Corps. With the creation of the Tennessee Valley Authority, the federal government launched a major ecological and social experiment in land use planning.

FDR also took a global perspective on conservation. In October 1944, in the depths of World War II, he sent a proposal to Secretary of State Cordell Hull for a meeting of all "the united and associated nations" in what would be "really the first step toward conservation and use of natural resources—i.e. a gathering for the purpose of a world-wide study of the whole subject." His memo added: "I am more and more convinced that conservation is a basis of permanent peace."[42] Roosevelt died the following spring, before his plan could be acted on. His successor, President Harry S. Truman, wrote to the head of the U.S. delegations to the United Nations' Economic and Social Council in 1946, and ECOSOC did hold a conference on resource conservation in 1949. But as Lynton Caldwell noted, that meeting was limited to an exchange of technical information and "had few tangible results."[43]

In the first half of this century, there were some scattered additions to the still minuscule body of international environmental law, primarily in its fur and feathers aspects, as well as the creation of institutions for the protection of nature. In 1909 the first International Congress for the Protection of Nature assembled in Paris with delegates from all over Europe. Out of that meeting arose the Consultative Commission for the International Protection of Nature, which numbered seventeen European nations and gave itself the task of collecting, classifying, and publishing "every item dealing with the international protection of nature."[44] The commission "set a precedent as being the first intergovernmental agency dealing with the protection of nature."[45] The outbreak of World War I, however, prevented the commission from carrying out its mission.

The 1911 Convention between the United States and Other Powers Providing for the Preservation and Protection of Fur Seals gave a measure of protection to a heavily exploited marine resource. The African Convention Relative to the Preservation of Flora and Fauna in Their Natural State, signed in 1933, was a noteworthy advance in protecting African wildlife from excessive hunting and harassment.[46] An Arbitration Tribunal finding in 1933 that assigned damages to a Canadian smelter for pollution causing harm in the United States was a landmark in fixing legal responsibility for transboundary pollution. But it was long an exception rather than the rule in such disputes over pollution.[47]

. . .

A number of private initiatives in multinational environmental activism were taken in the earlier part of the century. Notably, the International Council for Bird Preservation, "the oldest of the international wildlife conservation organizations," was formed in 1922 to coordinate the protection of migratory birds.[48] A Dutch conservationist, P. G. Van Tienhoven, working through the International Union of Biological Sciences, created the International Office for the Protection of Nature in 1934.[49]

For a long time neither governments nor private efforts slowed, even slightly, a careening, heedless industrial civilization to lessen the toll it was taking on the natural world and on the bodies and spirit of humans themselves. The great majority of people simply did not understand and therefore did not care about what was happening to their habitat. In the early part of this century there were dramatic breakthroughs in protecting public health through sanitation, hygiene, and advances in medicine. But in the absence of public pressure, governments around the world gave little or no attention to the waste and despoliation of resources, the blight of pollution, or the loss of natural beauty.

In the first half of the twentieth century, however, a deep intellectual foundation was laid for achieving a clearer understanding of humanity's evolving relationship with the natural world and of the consequences of that relationship. From science, economics and the other social sciences, ethics, and even religion emerged a new, ecolate worldview that, by the end of the century, challenged the long-prevailing values of uncontrolled scientific development, unfettered industrialism, unrestrained market competition, and ever expanding material consumption as the goal of modern society and human aspiration.

. . .

Early in the century a substantial conceptual breakthrough regarding the evolutionary effects of human activity was made by a Russian mineralogist, Vladimir I. Vernadsky. In his great work, *La Biosphere,* published in 1929, Vernadsky explained that we inhabit a world that is significantly the creation of human thought and activity. He divided the world into three spheres. One is the "geosphere," the geological realm of minerals, rocks, liquids, and gases. A second is the "biosphere," the thin layer of air, water, plants, animals, and microorganisms that, interacting with the geosphere and absorbing energy from the sun, is the basis of life on earth.[50] The third realm, the realm of human thought, communication, and action, which increasingly manages, changes, and could damage the biosphere, Vernadsky called the "noosphere," from the Greek word *nöos*, which means "mind" or "intellect." The noosphere, Vernadsky wrote, "is a new geological phenomenon on our planet. In it for the first time man becomes a *large scale geologic force*. He can and must rebuild the providence of his life by his work and thought, rebuild it radically in comparison with the past."[52]

The inescapable logic of Vernadsky's concept of the planet is that humans, to survive in the world they were now able to shape, must take responsibility for preserving the biosphere by assuring that their economies, technologies, and political arrangements operated in harmony with the physical environment.

This basic theme of the need for harmony between humans and nature and its variants was taken up by a growing number of scholars, scientists, and even theologians as the century progressed. The French Jesuit and paleontologist, Pierre Teilhard de Chardin, who also used the noosphere concept to define the created world of human thought, criticized the Christian doctrine that nature had been static since the creation and insisted that it was a dynamic, creative force. Humanity, he said, would save itself not by winning free of nature but by working with and becoming one with nature.[53] The notion that respect for and care of nature was an ethical duty also gained currency in the earlier years of this century. Albert Schweitzer, the physician, musician, and humanist, concluded that the position of power that humans had attained in the natural world did not give them a right to exploit nature but a responsibility to protect it. The ethical person extends the same "reverence for life" to all life that he gives to his own. The ethical person, Schweizer wrote in 1923 in a passage that recalls Eastern Jainism, "shatters no ice crystal that sparkles in the sun, tears no leaf from its tree, breaks off no flower, and is careful not to crush any insect as he walks."[54]

The growing power and destructiveness of twentieth-century technology—the overwhelming intrusion of the "machine in the garden," to use historian Leo Marx's metaphor—and the rapid encroachment of the cities into the rural landscape generated worried reappraisals of humanity's relationship with nature. Lewis Mumford, taking up the ideas of the Scotsman Patrick Geddes, who coined the word *megalopolis* early in the century, warned that Western civilization had taken a wrong turn since the science of Bacon, Newton, and Galileo reduced the universe to a set of discrete parts and principles. This mechanistic view of the world had led Western civilization to conclude that it could subjugate and rule nature with science and technology. As a result, modern civilization was being overwhelmed by the megalopolis and the "megamachine," artificial creations that enslaved humans rather than, as intended, serving them.[55]

The internal logic of the science of ecology, resting on the premise that all parts of the community of life are interdependent, seemed inevitably to point to a merger with human ethics. Although anticipated by many thinkers, including Darwin, Thoreau, Muir, and Schweitzer, that merger was long in being consummated. For many, however, the gap between science and ethics was closed by the American ecologist Aldo Leopold in his book *A Sand County Almanac*. In that book, Leopold proposed a new "land ethnic," based on science but also on love and reverence for the natural world. Humans, he noted, are but part of the "pyramid of life."[56] But they had acquired the power to destroy other parts of the pyramid and by so doing threatened the entire structure, including the human community. In thinking about uses of the land, it is necessary to think scientifically, aesthetically, and ethically, not just in terms of economic benefit, Leopold warned. "The land ethic," he explained, "simply enlarges the boundaries of the community to include soils, waters, plants, and animals, or collectively: the land." Such an ethic "changes the role of *homo sapiens* from conqueror of the land-community to plain member and citizen of it."[57]

Leopold's land ethic helped broaden the horizons of the young conservation movement in the United States, and its amalgamation of science and ethics eventually rippled across the world. It entered the canon of the slowly evolving environmental impulse that was opening the eyes of many Americans and Europeans to the toll that their increasingly affluent consumer societies were taking on nature and the quality of life. But the number and influence of those who watched and worried about the destruction of nature and the threat to humans were small. And for much of the first half of this century even those few did not realize the true magnitude of the dangers created by human power.

. . .

All of that changed in a sudden brilliant burst of light—a light brighter than the sun and of a kind never before seen on earth—that changed night into day in the early hours of July 16, 1945. The road of dark and riddling knowledge on which Prometheus led mankind had ended in a desert outside Alamogordo, New Mexico. Watching the pillar of fire ascend from the first atomic explosion that fateful morning, J. Robert Oppenheimer, the physicist who played a key role in the Manhattan Project, thought of a line from the *Bhagavad Gita:* "I am become death, the shatterer of worlds." Weeks later, on August 6, death descended on Hiroshima, as the *Enola Gay* dropped the first atomic bomb, leaving seventy-eight thousand dead or dying children, women, and men behind as it flew away from the instantly shattered city.

With the bomb, as environmental historian Donald Worster has noted, "For the first time in some two million years of human history, there existed a force capable of destroying the entire fabric of life on the planet." Worster believes that "the age of Ecology" began at Alamogordo.[58] What certainly did happen on that desert was the end of humanity's innocence about its own power. In that unearthly light, it was impossible not to see that we had achieved a savage dominion over nature. The meaning of Promethean fire had, at last, become plain.

. . .

In the post–World War II era, technologies less apocalyptic but, in their ubiquity and daily use, effectively more destructive than the bomb transformed the landscape and the physical, chemical, and biological systems that sustain life at an accelerating rate. Powerful machines, such as the bulldozer and the chain saw, permitted humans to shape the land to their desires with relative ease. The automobile and truck enabled humans to live their daily lives over widening areas and the cities to dominate and increasingly eliminate the countryside. Synthetic chemicals produced for industry and agriculture flooded into the air, water, and soil. Rising affluence in the industrial societies of North America, western Europe, and Japan generated an explosion in the production of consumer goods, a skyrocketing demand for primary products to make them, effluents from the production processes that choked the air and water, and mountain ranges of solid and hazardous waste. In the poorer countries of Latin America, Africa, and Asia, meeting the needs of soaring populations, exploding economies, and the demands of export markets in the North placed untenable burdens on the soil, water, forests, and other resources. Increasingly, control of the earth's resources passed from the hands of individuals, families, and even sovereign governments to impersonal transnational corporations accountable only to their own stakeholders. According to one well-informed estimate, fully half of the total transformation of the earth at the hands of humanity has occurred since World War II.[59] We were badly fouling our own nest, but as former U.S. environmental official Milton Russell observed, "the stench was overpowered by the stronger perfume of money."[60]

But the assault on the environment and the threat it posed to human welfare was no longer going unnoticed. A growing cadre of scientists, scholars, social activists, writers, lawyers, government officials, and even business executives, in the United States but in many other parts of the world as well, were joining the environmentalists in raising their voices in protest and warning. They also demanded and, with increasing frequency, achieved change.

The first major postwar environmental victory was the result of increasingly disturbing evidence that radioactive fallout from the testing of nuclear bombs in the atmosphere was passing into the food chain and, through mothers' milk, into the bones and teeth of babies. A campaign by scientists, initially led by Linus Pauling and Herman J. Muller and coordinated by Barry Commoner and the St. Louis Committee for Nuclear Information, was a major factor in the eventual negotiation of the 1963 Nuclear Test Ban Treaty.

Commoner, a biologist, socialist, and political activist, remained for many decades an eloquent Cassandra of environmental disaster and a gadfly for change. In his best-selling book *The Closing Circle* he warned that humans had "broken out of the circle of nature" with the means they had devised to gain wealth by conquering nature. "The end result is the environmental crisis, a crisis of survival. We must learn how to restore to nature the wealth that we borrow from it."[61]

In the postwar years a number of eloquent neo-Malthusians, including Fairfield Osborn, Garrett Hardin, and Paul Ehrlich, raised the alarm over erupting human numbers and their economic activities and the potential harm that such growth was likely to do to the ecological systems and human society. Internationally, the most influential neo-Malthusian voice was that of the Club of Rome, a loose association of economists, educators, scientists, and industrialists from twenty-five countries assembled by the Italian management consultant Dr. Aurelio Peccei. A report of the club's Project on the Predicament of Mankind, titled *The Limits to Growth*, generated enormous interest—and controversy—around the world. Published in 1972 and based on computerized projections of demographic, industrial, social, and ecological trends, the report found cause for both deep concern and hope. Its first conclusion was that "if the present growth trends in world population, industrialization pollution, food production and resource depletion continue unchanged, the limits to growth on this planet will be reached sometime within the next one hundred years. The most probable result will be a rather sudden and uncontrollable decline in both population and industrial capacity." But the report also concluded that it was possible to alter the trends that were leading down that dark road if the people of the world decided to change direction.[62] The executive committee of the club, led by Dr. Peccei, commented that the report demonstrated that "concerted international measures and joint long-term planning will be necessary on a scale and scope without precedent."[63]

Limits of Growth touched off a fierce debate between the "catastrophists," who demanded quick, concerted action to deal with the ecological threats endangering the planet, and the "cornucopians," who contended that the threats were greatly exaggerated and urged that the nations of the world continue with business as usual.[64]

One of the most potent catalysts in mobilizing individual, national, and international reappraisal of the impact of industrial civilization on the natural world and the threat it poses to human and nonhuman life was a book: the 1962 publication of *Silent Spring* by Rachel Carson.

Carson, a marine biologist with the U.S. Fish and Wildlife Service and a best-selling author, wrote *Silent Spring* to demonstrate that chemical poisons and other industrial and agricultural technologies were placing the entire chain of life, up to and including humans, in grave peril. She dedicated her book "To Albert Schweitzer who said 'Man has lost the capacity to foresee and to forestall. He will end up by destroying the earth.' "[65] Focusing on the scientific evidence of the effects DDT and other synthetic chemicals were having on the natural world, her book was, in fact, a moving polemic against the arrogance of human efforts to control nature. "The most alarming of all man's assaults upon the environment," she wrote, "is his contamination of air, earth, rivers and sea with dangerous and even lethal materials. This pollution is for the most part irrecoverable; the chain of evil it initiates not only in the world that must support life but in living tissues is for the most part irreversible." The spreading contamination, she warned, is "changing the very nature of the world—the very nature of its life."[66]

Silent Spring was a call to arms. And it was heeded by peoples and governments, first in the United States and then around the world. Earth Day, April 22, 1970, when millions of Americans demonstrated on streets and campuses to protest the abuse of the natural world, marked the emergence of environmentalism as a mass social movement in the United States. The U.S. government was quick to respond, creating a flood of landmark environmental statutes and institutions to carry out the new laws. Other nations were not far behind. A study of the formation of "the new environmental consciousness" in northern Europe noted that "as in other countries, in Sweden it was Carson's book that served to usher in the modern era of environmentalism."[67] Where there had been virtually no environmental agencies in existence, in the years after Carson's book most countries of the world had installed them (although in many of those countries the agencies were powerless). Nongovernmental organizations dedicated to protecting and preserving the environment also proliferated.

At first the new environmentalism emerged largely at the local and national levels. But the realities of depleting stocks of resources, transboundary pollution, the economic and social unrest and mass migrations caused by degraded ecological systems, and the rising tide of public alarm around the world very quickly forced national governments to raise their eyes beyond their own borders.

Some saw in the need of nations to face these common problems a new hope for uniting a badly divided world.

Then UN secretary-general U Thant expressed that hope in 1970, "that in saving ourselves by preserving our environment we might also find a new solidarity and a new spirit among the governments and peoples of the earth."[68]

That aspiration is still largely unrealized. But the world has moved forward over the past quarter of a century.

Chapter 3

Springtime in Stockholm

Let sea-discoverers to new worlds have gone,
Let maps to other, worlds on worlds have shown,
Let us possess one world, each hath one and is one.
—John Donne, *The Good Morrow*

On March 21, 1946, history began a short visit to the North Bronx. On a cool, cloudy, half-spring, half-winter day, the General Assembly of the new United Nations Organization convened in the faux-Gothic gymnasium building of Hunter College in the Bronx, its first temporary New York City headquarters. As trains of the Jerome Avenue "El" roared deafeningly over their heads, workers, merchants, mothers, and children from the lower-middle-class neighborhood of well-kept apartment buildings and small shops pressed against the chain-link fence surrounding the college campus to watch the limousines and taxis drive up and to catch a glimpse of the delegates arriving from around the world.

A half-century later the sense of excitement, hope, and joy with which the UN was welcomed is still vivid in the memory of one who was a young boy standing at that fence. A long, murderous, tragic war had just ended with the utter defeat, so it seemed, of the forces of evil. A new era of peace and plenty was clearly visible on the horizon. The UN would be the instrument through which the age-old dream of brotherhood among nations and justice for all of the earth's people would finally be realized. Or so we were told by our leaders, and so we believed. Two months earlier, at the General Assembly's first meeting, held in London, British Prime Minister Clement Atlee eloquently outlined the "ultimate aim" of the UN: "It is not just the negation of war but the creation of a world of security and freedom, of a world which is governed by justice and the moral law. We desire to assert the pre-eminence of right over might and the general good against selfish and sectional aims."[1]

But the high expectations for the UN quickly came crashing down, even, in fact, before it moved from the Bronx to Lake Success on Long Island. Almost immediately, Soviet ambassador Andre Gromyko stalked out of a meeting of the Security Council over the issue of Russian troops in Iran.[2] The forum for peace became and remained a battlefield of the long Cold War between the United States and the Soviet Union. The swelling number of newly independent nations joining the UN, many of them belonging to a bloc that would be called the Third World, increasingly used the General Assembly and the agencies of the UN organization to vent their frustration, anger, and resentment over their poverty and their powerlessness. Although it engaged in several successful peacekeeping operations, the UN could not prevent wars in the Middle East, in Korea and Indochina, between China and India, in Cyprus, in Afghanistan, in Angola, in the South Atlantic, and along borders in various other parts of the world. It could not stop the bloody internal struggles in many of the new nations nor ease the repressions of totalitarian regimes against their own peoples. As one observer noted, on many of its issues, "the U.N. does not appear to be bigger than the sum of its parts, and failure to make real headway rests with its member states. The U.N. has often been used by them to pursue their own self-interest, rather than in support of international interests."[3]

Although the World Bank and other Bretton Woods multilateral financial institutions created as part of the UN system after World War II did try to prime the development pump in Africa, Asia, and Latin America, over the long run they were unable to make much progress in closing the enormous gap between the rich and poor countries of the world. In practice, the free-market, free-trade approach on which postwar development programs were largely based worked to the disadvantage of the struggling new nations, which could not compete with the overwhelming economic power of the industrial North.[4]

And while the UN system produced many able and dedicated international public servants, parts of the system gradually hardened into ossified bureaucracies with permanent secretariats jealously guarding their own turf and perquisites.

Inevitably, much of the early hope for the UN faded and was replaced by disappointment and cynicism. Many wrote it off as just another stage for acting out the age-old, endless power struggles. A number of governments, including the United States, fell far behind in meeting their financial obligations to the system. Sovereign nations frequently bypassed the organization in the conduct of their multilateral diplomacy.

But if the UN functioned all too often as a forum for airing hostilities and serving the narrow self-interest of nations or blocs of nations, it also provided the people of the earth something they had never had in history: potentially workable international institutional machinery for helping each other address their common problems and coordinating efforts to meet common needs and solve common evils afflicting the human condition. Arms of the UN worked with varying degrees of efficiency to promote economic development, human rights, the rescue of refugees, the welfare of children, fair labor standards, education, health, food production, better telecommunications, aviation, international shipping, the improvement of human habitats, and the advancement of science. In time the UN also became the chief mechanism for international cooperation in protecting the global environment. It would take several decades.

. . .

The explosion in population and industrial activity of the post–World War II years brought with it a quantum leap in pollution and strains on the earth's resource base. New technologies, including high-compression internal combustion engines, synthetic chemicals, and nuclear energy posed severe if unstudied risks to human health. Consumption and waste soared in the rich industrialized countries while the many of the poorer societies suffered from the ill effects of industrialization without enjoying many of its benefits. Dramatic and tragic environmental disasters, including the destructive breakup of the oil tanker *Torrey Canyon* in the North Sea in 1967 and the terrible outbreak in the 1950s of Minamata disease among Japanese coastal dwellers who had eaten seafood contaminated with mercury, aroused a global public to the damage human activity was causing to the chain of life. The spread of the cities into the surrounding countryside sent people in many regions into mourning over the loss of nature. In some countries, particularly in the United States, citizens who had mobilized for peace, for equality among the races, for women's rights now turned their anger and activism to environmental issues, forming and joining many new nongovernmental advocacy groups. Pictures of a small and lovely planet Earth sent back by the first humans to venture into space made many people grasp for the first time the vulnerability of their only habitat.

But if demographic, economic, social, and political forces eventually weighed in heavily, it was, in large measure, the involvement of the UN in scientific issues that first led the international organization on a path that would eventually give it the central role in protecting the global environment. Early UN scientific conferences, such as the 1949 Scientific Conference on the Conservation and Utilization of Resources, failed to address the threat that human activity was imposing on natural systems.[5] But in 1957 the International Geophysical Year demonstrated the value of international cooperation on research scientific issues affecting the planet as a whole.[6] Scientists in several countries called for an international meeting on the human impact on the biological world. Politically vocal scientists, including Barry Commoner in the United States and Andre Sakharov in the Soviet Union, were making urgent pleas for serious attention to be paid to threats to crucial ecosystems. Then, in

September 1968, the International Conference of Experts for Rational Use and Conservation of the Biosphere met in Paris under the sponsorship of the UN Educational, Scientific and Cultural Organization.

The Biosphere Conference was the first major international scientific gathering to look at the overall human impact on the conditions necessary for life on earth. On its agenda were air and water pollution and the deterioration of wetlands, forests, rangeland, and other natural resources. There was consensus that changes in many natural systems were reaching critical thresholds. The conference attendees noted a demand for changes in the way the biosphere was being used by humans, and they called for expanded international research into the interrelated problems that were causing biological decline. They discussed the relationship between economic development and ecological decline.[7] Out of the conference emerged the ongoing "Man in the Biosphere" program, which advanced global scientific cooperation in addressing concrete problems arising from the interaction of humans and their biological habitat.

Lynton Caldwell has remarked that the twenty years between the 1949 UN Scientific Conference and the 1968 Biosphere Conference witnessed "as fundamental a change in perceptions of international responsibilities for the global environment as any that have occurred since the establishment of permanent international organizations."[8] The change, however, was one of scientific perception, not of political will by national governments or of concrete steps by the international community to address the perceived problems.

A political breakthrough came four years later.

. . .

The UN Conference on the Human Environment, held in June 1972 in Stockholm, had its most immediate roots in a regional environmental problem afflicting northern Europe. Leaders in Sweden and Norway were becoming increasingly alarmed about the impact of growing pollution, particularly acid rain, drifting over their lands from industrial emissions in Britain and central Europe. The pollution was acidifying their lakes and damaging their forests and cropland. While the causes and effects of acid rain had been described a hundred years earlier by the British chemist Robert Angus Smith and facts about its wide dispersion were well known by the middle of the twentieth century,[9] there had been little national, much less international, action to deal with the problem. But the Scandinavians, particularly the Swedes, who had taken activist, supportive roles in the UN since its formation, now saw the UN as an instrument for addressing acid rain and other international environmental problems. Sverker Astrom, Sweden's ambassador to the UN pressed the secretariat to convene an environmental conference; and on December 6, 1968, the General Assembly, acting on a resolution of the Economic and Social Council, authorized the Human Environment conference.[10] Because of the role played by the Swedes, the meeting would be held in Stockholm.

The General Assembly did not originally intend that the conference would actually *do* anything other than talk about the problems. Its resolution said that the conference had been called because it was "desirable to provide a framework for comprehensive consideration within the United Nations of the problems of human environment in order to focus the attention of Governments and public option on the importance and urgency of this question and also to identify those aspects of it that can only or best be solved through international cooperation." It took two years of reconsideration and negotiation for the General Assembly to give the conference an action-oriented mandate, to actually attempt to adopt new standards and treaties to confront the mounting global environmental crisis.[11] And it took the appointment of a young, able, and inspiring but at that time somewhat obscure Canadian foreign service official as secretary-general of the conference to assure that something significant would indeed take place in Stockholm.

. . .

Maurice Frederick Strong was Canada's deputy minister of external affairs for foreign aid when he was put in charge of the UN environment conference. His route to that job had been improbable. In fact, Strong's adventurous "rags to riches to international achievement and renown" life story would have caused Horatio Alger himself to lift a skeptical eyebrow.[12]

Strong grew up in extreme poverty on central Canada's austere and windy prairie. He was born April 29, 1929, in Oak Lake, Manitoba, a town on the main line of the Canadian Pacific railway. His father was a young assistant station agent for the railroad with a seemingly promising career in front of him. His mother unusually for her time, had a university education and had worked as a journalist and teacher. But when the Depression struck Canada, the elder Strong was laid off his job. He first became a laborer, working on maintenance of the line, until he lost that job, too, as the railroad continued to retrench. For many years thereafter, he was unable to find a full-time job. He did not receive a regular paycheck again until nearly a decade later, when World War II began and he entered the Canadian armed forces. Until then he traded his labor to farmers and merchants for food and firewood. The family kept a garden and put up vegetables for winter. Young Maurice went into the fields with his mother to pick berries. A kindly local storekeeper kept extending the family's credit so that they could buy a few groceries.

The Depression was a period of great adversity for the Strongs. Occasionally, the family was reduced to eating dandelions and pigweed.[13] There were times, Maurice recalled, when his father had to go out in the snow to cut wood without adequate shoes; he would bind rags around his feet to try to keep them warm. His mother, not used to hardship, suffered a nervous breakdown, was in and out of mental institutions for the rest of her life, and died while only in her fifties. "The Depression literally destroyed her," Strong would say many years later.

Out of the Depression years, Strong carried a burning awareness of social injustice in the world and an ardent sense of personal social responsibility that determined much of his later career.

> Our house was right beside the railroad station. Some of my earliest memories are of trains going both ways loaded with people—literally hundreds of people on those freight trains—searching for work, for something. Because our house was right by the tracks, when the trains stopped they would come over and ask for food and water. Of course we were very poor, but we always gave them what we could. Very early in the game I saw that something was very wrong. People wanted to work and they needed things but couldn't get them.

He felt the sting of the world's unfairness directly: "Almost everybody was poor, but because we were poorer than most, some families wouldn't let their kids play with me."

Despite all the hardship, young Strong flourished intellectually. His mother, in her lucid periods, would read to him from Shakespeare and other classics. Encouraged by his high school principal to excel, he skipped grades and graduated at age thirteen with classmates who were seventeen and eighteen. He also won a financial scholarship to help him go to college. When the check arrived, however, he took it straight to the local grocery store to pay off his family's long-standing debt. He told himself he would find some other way to attend a university. He never did.

In 1943, at the age of fourteen, Maurice Strong ran away from home. He had intended to join the armed forces but was rejected because of his age. So he hitchhiked and jumped a freight to Thunder Bay on Lake Superior, determined to get a job on one of the lake liners. He was on the train for three days with nothing to eat except some carrots he pulled from a garden. When he got to the town of Thunder Bay, exhausted and half-starved, he went into a restaurant and ordered a full meal, knowing that he did not have a penny in his pocket. When he finished, he asked to see the owner of the restaurant and told him he couldn't pay for his meal but would be happy to work off his bill. The owner told him not to worry about it and to come back any time he was hungry. He then slept on the doorstep of a school and the next morning went aboard a lake liner owned by Canada Steamship Lines and asked the second steward for a job. The steward said that he didn't have time to talk because the boat was about to leave. "So I hid in a closet," Strong recounted. "When I knew from the sound that the boat was already out, I reappeared and went to look for the steward and said, 'I'm really sorry I didn't get off the boat.' Of course, he gave me hell, but he also gave me a job. Anyway, that's how I started. Later on, I bought that shipping company."

Still wanting to get into the war, Strong hopped another freight train when the lake voyage was finished, this time heading toward Vancouver and a possible berth in the merchant marine. He was riding a flat car and was covered with coal dust and drenched with rain when the train stopped in Oak Lake. "My home was right across the way. I was really tempted to go home. But I felt that would be going home defeated. So I continued on." One day on this journey he picked up a newspaper that had a story about Roosevelt and Churchill agreeing to create a new international organization called the United Nations after the war ended. "For some reason I developed a fixation about the U.N. I had no education or anything, but somehow I thought I was going to go to the United Nations."

He did get a job on a merchant marine vessel carrying troops to Alaska. Along the way he wrote to his father to tell him where he was. A few days later, while waiting to leave with a ship going to the Aleutian Islands, two policemen picked him up at his father's request and took him to a detention home. "They called it a detention home, but it was really a jail for younger people." A short while later his father showed up and asked him to come home for a year because his mother was worried. So he did.

Strong's youthful adventures are related at some length here because they show the initiative, overreaching, and risk taking, as well as the ability to charm people into helping him, that would mark a career of remarkable success in business and diplomacy and eventually as the undisputed Mr. Environment of geopolitics.

After somewhat less than a year back home, he answered a want ad for a job in the Arctic with a Hudson Bay trading company, beginning an astonishing ascent in the world of business. In the Arctic, his love of the out doors and of wildlife, acquired while roaming the prairie and hills around his hometown, blossomed as he spent time learning the ways of the local Inuit people. He also spent much of his spare time collecting minerals. An American prospector, passing through the post and impressed by Strong's self-taught knowledge of minerals, said he would give him a job if he came to Toronto. When Strong arrived, however, no job was available, so he found work as a delivery boy for a stationery business. Soon thereafter, however, the American formed a prospecting company and asked Strong to join as a partner. One day, a visiting friend of the American's Canadian wife, and official of the UN, learned of Strong's still intense interest in the U.N. and told him he would help him get a job in the international organization. Strong leaped at the chance and journeyed to Lake Success, where he was give a clerical position in the security office. He was fascinated by the UN and met many of its notable personalities, including Gromyko and Soviet foreign minister Vyacheslav Molotov. But he soon realized that, without education or connections, he would never rise far in the hierarchy, and he left for a brief stint in the Canadian armed forces, vowing, however, to come back in a high position.

He returned to Canada in 1947 and took an apprentice position with one of Canada's largest investment houses, where he made himself an expert on the oil and gas industry. Armed with that knowledge, he fast-talked the head of the firm into sending him as an oil analyst to its Calgry branch, despite skepticism because of his youth. Soon he was hired away by the new Dome Petroleum Company and quickly became its vice president for finance. By age twenty-three he had gotten married, made a small fortune, decided business was not what he wanted, and resigned his executive position to take a round-the-world trip with his wife. He interrupted his travels in Kenya, then in the middle of the Mau Mau struggle for independence, and was intrigued both by the politics of this emerging African nation and by the East African wildlife. He took a job as assistant to the general manager of the Caltex company's service station operations in East Africa. Then, becoming interested in the development problems of the Africans, he decided to return to Canada to work in his government's foreign assistance agency. But he was turned down because of his lack of a college degree. He was rejected for a position with Canada's YMCA for the same reason.

"I was really at the low point in my life. Then one night I had an insight. Business was not what I meant to do but business seems to be the only place I have the ability to get anywhere. So I should go back into business because that could be my platform to power." After a brief return to Dome Petroleum he formed his own management firm and soon took over a small, bankrupt oil company, which became the multimillion-dollar Norcen Company; he made another modest fortune and was named president of the Power Corporation of Canada. At age twenty-nine he was chief executive of one of his country's biggest, most powerful business organizations.

Six years later, Canada's prime minister, Lester Pearson, offered Strong the post of deputy foreign minister in charge of external aid. At lunch with Pearson, Strong mentioned that only a few years earlier he had applied for a lower-level position with the aid office and had been rejected because of his lack of a college degree. "Pearson laughed and said, You're lucky. If you had the qualifications, you would have gotten the job and, no matter how well you performed, you wouldn't at this stage of your life be beyond the lower middle levels of the department, and I couldn't possibly be offering you the job of heading it.' "

As deputy minister, Strong created the Canadian International Development Agency and became an actor on the world stage, a role he had long desired.[14]

Because of his interest in the UN he began attending meetings to which a deputy minister might normally have sent subordinates. He became active in World Bank activities and generally began establishing a presence and making a name for himself in development assistance circles.

By 1970 preparations for the 1972 Stockholm conference had been underway for nearly a year and a half. They were not going well. The Swiss professor in charge of the conference secretariat apparently had not given the preparations a sense of direction or ultimate purpose. The Swedes, who had pushed hard for the meeting, were deeply concerned. Sverker Astrom, supported by Sweden's prime minister, Olaf Palme, sounded out Strong to find out if he was interested in taking over. He was. In addition to his enthusiasm for the UN, he cared about the environment. "As I got into business, it was impossible not to realize that resource development has a major impact on nature and the environment. I was deeply interested in the developing countries. So I saw the environment issue as a new way of coming after development issues."

The Swedes and their allies quickly convinced Secretary-General U Thant and the General Assembly that Strong was the man for Stockholm. It was somewhat harder to persuade Canada's prime minister, Pierre Elliot Trudeau, to let Strong go, but he eventually assented. In 1970, at the age of forty-one, Strong became secretary-general of the UN Conference on the Human Environment and Under Secretary of the UN, the organization in which, a relatively few years earlier, he could not rise above a clerical position.

Strong jumped into his new job with his characteristic and indefatigable energy, improvisational skills, and ability to enlist the assistance of talented, accomplished people from around the world. He persuaded the scientist René Dubos and the economist Barbara Ward, both brilliant writers, to produce a book, with input from dozens of experts, that would set the stage for Stockholm. The result, *Only One Earth,* sounded an urgent alarm about the impact of human activity on the biosphere but also expressed optimism that a shared concern for the future of the planet could lead humankind to put aside ancient conflicts in the face of common danger. The book became a best-seller.

Strong actually took over the secretariat in 1971 and faced a herculean task in completing the arrangements in a little more than a year, while simultaneously having to leap more than the usual number of political hurdles. The Soviet Union, while taking an active part in the preparations, eventually boycotted the conference itself because, under the UN's Cold War rules, East Germany was not allowed to participate. Most of the other Eastern European nations also declined to attend. The Chinese Communist government did send a delegation to Stockholm—it was Beijing's first major UN conference—but because of its strident complaints about the supposed imperialist goals of the industrialized countries, its presence was viewed by many, including environmentalists, as a mixed blessing. Most of the developing countries considered environmental protection a luxury that only the rich nations could afford[15] and voiced the suspicion that the conference was yet another ploy by the rich men's club to keep the poor countries from industrializing.[16]

In preparing for the conference, the secretariat accumulated well over twenty thousand pages of documentation, which it eventually managed to boil down to eight hundred pages.[17] Perhaps the document with the single greatest effect on the conference was the Founex report, prepared by a panel of experts (appointed by Strong) who met in Founex, Switzerland, in June 1971. The panel of distinguished scholars, economists, and other experts forged a vital link between environment and development. Its report noted that while the existing concern about the environment sprang from the production and consumption patterns of the industrialized countries, much of the environmental problems in the rest of the world were a result of underdevelopment and poverty. The report called for the integration of environment and development strategies and urged the richer nations to provide the additional development assistance to help the poorer countries achieve this goal. "If the concern for human environment reinforces the commitment to development," the report asserted, "it must also reinforce the commitment to international aid."[18]

Strong believes that the Founex report gave legitimacy to the notion that environment and development are interdependent and that the report also was a factor in persuading many of the developing countries to attend the Stockholm conference.[19] But twenty years later the debate over environment and economics would still be unresolved.

. . .

"On a beautiful spring mourning in Stockholm . . . the world community embarked on an extraordinary journey of hope," Strong recalled twenty years later.[20] From the start, however, it was a bumpy journey. The representatives of 113 nations gathered in the Stockholm Opera House under the conference theme of "Only One Earth.[21] But there was so much bickering, rhetoric about protecting national sovereignty, and disagreement about the purpose and goals of the conference, that a conference newspaper put out by two nongovernmental organizations—Friends of the Earth and *Ecologist* magazine—ran a headline proclaiming "Only 113 Earths."[22]

In the absence of the Soviet bloc the key political dynamic of the conference was the sharp differences between the North and the South, the industrialized and the developing countries. For the rich countries the conference was about dealing with such issues as pollution, resource depletion, and population pressures, as well as finding technological solutions to many of these problems. The poorer countries, led by Brazil, India, and China, were deeply suspicious of the North's agenda, particularly about what they regarded as attempts to curtail their hard-won sovereignty.

Environmentalists and other nongovernmental organizations held their own unofficial forum in Stockholm on land lent by the Swedish government and dubbed "the Hog Farm." The unofficial gathering, which was invigorated by spirited presentations by the anthropologist Margaret Mead, Barbara Ward, Barry Commoner, and other well-known scientific and cultural figures, acted as a gadfly to the official conference. Many of the activists called for an end to international whaling, and a large rubber whale displayed at the Hog Farm was transformed by the media into a symbol of the conference. So effective were the activists on this issue that the conference agreed to propose a ten-year moratorium on whaling. Strong flew to the International Whaling Commission (IWC) meeting in London immediately after Stockholm to confront them with the conference's moratorium proposal. The IWC heard him, he later recounted, only because of the heavy pressure mounted by activists.

. . .

Stockholm was the first successful major foray by the environmentalists into the treacherous arena of international politics. The ability of the environmentalists and their allies in the nongovernmental community to arouse public opinion gave them a permanent role in the politics of international environmental policy-making after Stockholm. Their achievements contributed to the rise of nascent environmental movements around the world, including many of the poor countries.

Many of the environmental groups, reflecting the warnings of the Club of Rome report, argued for a new international order shaped by "steady-state," or no-growth, economics. Most of the industrialized countries, particularly the United States, argued for a business-as-usual approach of encouraging maximum economic expansion. The middle ground, expressed most forcefully by Strong, held that continued growth was imperative to alleviate poverty and allow the people of the world to achieve their material and cultural aspirations. But, he told the conference, "we must rethink our concepts of the basic purposes of growth. Surely we must see it in terms of enriching the lives and enlarging the opportunities for all mankind. And if this is so, it follows that it is the more wealthy societies—the privileged minority of mankind—which will have to make the most profound, even revolutionary, changes in attitudes and values."[23]

Stockholm was not a summit meeting, and other than Swedish prime minister Olaf Palme, the only head of state to attend was India's prime minister, Indira Gandhi. Mrs. Gandhi forcefully stated the developing world's position that "poverty is the greatest polluter," but she did so in a reasoned, conciliatory fashion and drew the conference's first standing ovation. "The inherent conflict," she said, "is not between conservation and development but between environment and the reckless exploitation of man and the earth in the name of efficiency." She called for a new view of the human condition centered on man "not as a statistic but an individual. The higher standard of living must be achieved without alienating people from their heritage and despoiling nature of its beauty, freshness and purity, so essential to our lives."[24]

The U.S. government, which had been a strong supporter of the conference and which had sent an able delegation led by Russell Train, the chairman of the first Council on Environmental Quality, and included William Ruckelshaus, the first administrator of the Environmental Protection Agency, nevertheless played a negative role in several respects. While the Nixon administration backed international cooperation to combat pollution and preserve marine and terrestrial resources, it generally opposed any change in existing economic relationships with the developing countries. Also, the United States at the time was still on the defensive because of the continuing war in Vietnam. Positions adopted by the United States were criticized by Europeans, as well as by Third World delegations, as "rigid" and "arrogant."[25] Nor did any other major country step forward to give the conference strong direction. As *New York Times* columnist Anthony Lewis wrote from Stockholm, the conference "needed a Thomas Jefferson who could lift the delegates above their parochial concerns and rally them behind a contemporary equivalent of the call for life, liberty and the pursuit of happiness. But . . . there is no Jefferson."[26]

Despite the absence of a Jefferson and the often sharp differences between the industrialized and the developing nations, the Stockholm conference was remarkably productive. In large part this was due to Strong's tireless diplomacy, to the prodding of the environmentalists, and to the demands of the Third World, not only for an expanded agenda, to include their economic issues, but an expanded definition of the environment, to include such issues as of poverty and the sharing of technology. One observer found that Strong's "missionary work" and his insistence that environmental protection and continued development were compatible did much to soften the North–South confrontation and permit the work of the conference to move forward.[27]

Perhaps the hardest-earned product of the conference was its final declaration, known subsequently as the Stockholm Declaration. A conference statement had been drafted by the preparatory committee before the conference began, but at the insistence of China and other developing countries it was renegotiated virtually in its entirety at the meeting itself to give more weight to the concerns of the nonindustrialized nations. The twenty-nine principles in the declaration are sweeping but general in nature, ranging from the imperative of preserving the earth's natural resources, ending the discharge of toxic substances, and preventing pollution to creating the economic and social conditions for an improved quality of life for all people. The declaration met the demands of the developing nations on such issues as fair prices for commodities. It also called for the destruction of all nuclear weapons and condemned apartheid, colonialism, and racism.

Stockholm was the first major international conference in which the government of the People's Republic of China participated. Its delegation took tough positions on many issues, but in the end a formula was worked out that enabled it to refrain from blocking consensus adoption of the Stockholm Declaration, a key to the successful conclusion of the conference.

The declaration, of course, did not carry the force of hard international law. Many of the delegations that voted it for it had no intention of following at least some of its principles, such as the elimination of nuclear weapons. But by accepting the declaration by a consensus vote,[28] the nations of the world did formally accept the principle that all people have a "fundamental right" to live in a environment "of a quality that permits a life of dignity and well-being," just as they have a right to freedom and equality. It was the first time that such a right was conceded by the sovereign nations of the world.

It was also the first time that those sovereign states acknowledged that all nations had a shared interest in protecting the global environment and a responsibility for domestic actions that affected the environment of other nations and the planet as a whole.

Analyzing the declaration a year later, Louis B. Sohn, a law professor at Harvard and authority on international law, noted that despite the lack of specificity, the overall "tone" of the document "is one of a strong sense of dedication to the idea of trying to establish the basic rules of international environmental law." He noted that the UN Declaration on Human Rights was originally viewed as no more than a "hortatory" document but that over the years it has had real world impact in strengthening those rights around the world. The Stockholm Declaration is likely to have a similar history, he wrote. "Having accepted the responsibility for the preservation and human environment, the international community will find in the Stockholm Declaration a source of strength for later, more specific action."[29]

As it turned out, Stockholm signaled the dawn of an era of international environmental lawmaking. Of the 140 multilateral environmental treaties that have entered into force since the 1920s, more than half were concluded since 1972.[30]

The conference also produced a 109-point "Action Plan" dealing with an incredibly broad range of issues. They included the environmental quality of human settlements; natural resource management; identification and management of pollution; the educational, social, and cultural aspects of environmental issues; environment and development; and organizational matters.[31] The plan included a number of ground-breaking proposals, including a recommendation that the developing countries be granted additional assistance to help them meet new environmental standards.

As with the declaration, the action plan was not binding on sovereign nations. But it stimulated a wave of activity by the nations of the world to begin to address the threats of pollution and resource depletion. In 1972 only a handful of rich countries had established environmental ministries, and those were generally brand-new. Ten years later more than one hundred governments had created such institutions, and by 1992 most of the nations of the world could boast environmental agencies.

To carry out their recommendations, the Stockholm delegations voted to create a fifty-four-nation Governing Council for Environmental Programs within the UN. They also called for a "voluntary fund" to support such functions as research into environmental issues, regional and global environmental monitoring and data collection, the exchange and dissemination of environmental information, and the promotion of environmentally beneficial technologies. Finally, the conference called for the creation of a "small secretariat" to serve as "focal point for environmental action and coordination within the United Nations system."[32]

The small secretariat evolved within a year into the UN Environment Program (UNEP). At the insistence of the African bloc, which wanted a UN agency on their continent, UNEP was headquartered in Nairobi, Kenya, instead of in New York or Geneva. Maurice Strong was named as its first executive director.

The birth of the UN environmental agency was probably the most significant concrete result to emerge from Stockholm. Political scientists Howard R. Alker Jr. and Peter M. Haas concluded that "the creation of a globally oriented United Nations Environment Program . . . was probably the most important institutional consequence of increased concern with global environmental change in the Cold War era."[33]

But Stockholm had a broader, more profound effect on the awareness of the peoples of world and their governments about the nature and breadth of the environmental challenge facing the planet. It demonstrated, Strong wrote a year after its conclusion, that "the environmental problem required a new view of man's relationship not only with the natural world, but with his fellow man. Involved was not only pollution, but a whole series of threatening imbalances in which the fate of rich and poor alike is joined." In fact, Strong added, the world owed thanks to the representatives of the poor at Stockholm for expanding "the world's environmental awareness to embrace not just the effluents and wastes, the dangers, disamenities and aesthetic insults found in industrialized societies, but the malnutrition, disease, illiteracy and human degradation which are the dominant characteristics of the human environment for most of mankind."[34]

Accounts written at the time of the conference suggest that it was marked by a rare unanimity of perspective among the disparate nations of the world in the face of common danger. A *New York Times* article remarked that "it was as though the nations comprising the family of man had become aware, as never before, of the vulnerability of their planet and how essential it is that they work in concert to preserve it."[35]

The notion that environmental problems could be managed by the concerted action of governments took somewhat tenuous root in Stockholm.[36] The conference gave environmental issues a legitimate and permanent place on the international diplomatic agenda. As Lynton Caldwell commented, the conference "is a major landmark in the effort of nations to collectively protect their life-support base on earth."[37]

Those issues, however, remained near the bottom of the geopolitical agenda for nearly another two decades. Many other problems related to the human condition in the natural world were largely ignored or inadequately addressed at the conference, including population growth, militarism, governmental corruption and mismanagement, the role and status of women, and oppression of indigenous peoples.

Stockholm was only a beginning. It was a single step—a major step, to be sure—on what would be and continues to be a long, uncertain journey toward the elusive goal of an effective international system for safeguarding a productive and beautiful planet and preserving it for the coming generations.

Chapter 4

The World As It Is

Between the idea
And the reality
Between the motion
And the act
Falls the Shadow.
—T. S. Eliot, *The Hollow Men*

A year after the human environment conference, Peter Stone, who had served as a senior information aide to Maurice Strong, published a book titled *Did We Save the Earth at Stockholm?* The book did not attempt to answer the question. There was no need to. The world, insofar as it needed saving from the human assault on the air, the land, and the water, on the other creatures with which we share the earth, and on one another—from hunger, from poverty, and from inequity between rich and poor and between the powerful and powerless—clearly was not saved by the actions of governments and individuals after the UN meeting.

Despite the new global awareness of the dangers, despite the linkages made between economic and ecological needs, despite the new institutional and financial mechanisms set in place, despite a wave of environmental treaties, and despite an active, expanding international environmental movement, there was little significant change in national or international behavior in the years following Stockholm. The pressing, often narrow realities of the immediate quickly renewed their grip on the affairs of governments, industries, and individuals.

Maurice Strong, looking back at the accomplishments of Stockholm two decades later, conceded that "nevertheless, the global environmental crisis continued."[1] He noted that many ecological, economic, and social trends had worsened rather than improved in the intervening twenty years. In the same year, Mostafa K. Tolba, who had succeeded Strong as executive director of the UN Environment Program, said when releasing a report, "The World Environment, 1972–1992," that "the state of the environment is worse than it was 20 years ago." Governments, he cautioned, must do more than sign international agreements to reverse the decline. "We need to change gears. We need a change of heart," he warned.[2]

The authors of *The Limits to Growth*, looking at environmental and economic trends twenty years after its publication in 1972, found that their projections that the world would exceed its carrying capacity in the absence corrective action were turning out to be all too accurate. *Beyond the Limits,* published in 1992, updated the computer model used for the earlier book and concluded that the rate at which human population and economic activity had expanded was physically unsustainable. "Human society has overshot its limits, for the same reasons that other overshoots occur. Changes are too fast. Signals are late, incomplete, distorted, ignored or denied. Momentum is great. Responses are slow." A course correction was still possible, but "we also believe that if a correction is not made, a collapse of some sort is not only possible but certain, and that it could occur within the lifetimes of many who are alive today."[3]

Beyond the Limits, like its predecessor, was criticized as a bit apocalyptic, even by some of those who worried about the fate of the earth. A number of Panglossian anti-Malthusians such as Julian Simon and Herman Kahn, free-market ideologues, growth-at-any-cost economists and development planners, and polluting and resource-consuming industries scoffed at environmental alarms as exaggerated and irrelevant. But a growing consensus of experts in and out of governments around the planet was arriving at the dismaying conclusion that human activity was making the world a more, not less, perilous place.

. . .

Not that those years had been without real progress in many areas. Certainly knowledge of ecological dangers created by human numbers and human activity—arcane subjects virtually unknown to most people—had penetrated deeply into the global consciousness. Just before the tenth anniversary of Stockholm, Noel Brown, the North American director of the UN Environmental Program (UNEP), asserted that "in 10 years, environmentalism has become a global value."[4] Environmental values also were being accepted by the poor nations of the world, which had vigorously opposed them in the past. UNEP's Tolba noted at that tenth anniversary that "it used to be that the developing countries said that environmentalism was a plot to keep them from developing. Now it is deeply embedded in their planning."[5]

The industrial nations, led in the 1970s by the United States, made substantial strides toward slowing domestic environmental decline and reversing the grosser symptoms of pollution and land degradation. In the United States the wave of environmental statutes passed by Congress and the institutions created at federal and state levels to enforce them did, in fact, produce improvements in air and water quality, in some areas substantial improvements. True, the progress often fell far short of the goals. William Ruckelshaus, in his second tour of duty as administrator of the U.S. Environmental Protection Agency (EPA) in the early 1980s, noted that most U.S. waters still had not reached the "swimmable, fishable" state called for by the Clean Water Act. "But at least they're no longer flammable," he added.[6] Ruckleshaus frequently pointed out how bad air and water quality would have become in the absence of the strenuous efforts led by the federal government in the 1970s. Intense efforts, often against strong resistance by industry, were made to reduce the flood of pesticides and other toxic substances into the nation's environment. Enactment of the Endangered Species Act imposed on the government the duty of protecting plant and animal species from extinction caused by human activity. New programs were undertaken to protect and rehabilitate wetlands, rangelands, farmland, and other parts of the American earth under assault by development. Membership in environmental groups at the local and national levels exploded as more and more Americans became aware of the growing dangers. Policymakers began, painfully slowly, to recognize the linkage between assaults on the environment and other symptoms of economic and social inequity in the United States.

Europe and Asia lagged behind the United States in addressing environmental problems but slowly began to catch up and, in some areas, move ahead. Japan made particular progress in dealing with the air pollution that had been choking and almost strangling Tokyo and some of its other large industrial cities. Green political movements, which will be discussed in a later chapter, forced governments otherwise single-mindedly fixated on economic growth and competition to start paying attention to the environmental price paid for the so-called economic miracles enjoyed by a number of western European countries in the post–World War II era.

It would be difficult, however, to make a case that substantial improvement was made in the environments of many of the developing countries in the wake of the Stockholm meeting. Indeed, the economic gains made by some of those countries in the 1970s was often at the cost of increased pollution, resource depletion, and the destruction of habitats of human populations and wildlife. The smog that blanketed rapidly industrializing Seoul, the traffic jams clogging the impoverished streets of Bangkok in the 1970s and early 1980s spring immediately to mind. But slowly and inadequately, many of those countries did begin to factor environmental concerns into their national economic plans. And the World Bank and bilateral aid donors began setting aside at least a small portion of funds for environmental projects, if often in ill-conceived and ineffective ways.[7]

Following the UN conference the pace of international diplomacy to address international and global environmental problems quickened perceptibly. Multilateral treaties and conventions experienced a high birthrate, although adherence to the treaties was frequently spotty at best. Among the major accords were the 1973 Convention on International Trade in Endangered Species of Flora and Fauna (CITES), the 1973 MARPOL Convention on the Prevention of Pollution by Ships, the 1975 Convention on Wetlands of International Importance, the 1979 Convention on Long-Range Transboundary Air Pollution, the 1982 UN Convention on the Law of the Sea, the 1985 Vienna Convention for the Ozone Layer and its 1987 Montreal Protocol, and the 1988 Convention on the Regulation of Antarctic Mineral Resources.[8] In addition, the International Whaling Commission finally agreed in 1982 to impose a moratorium on all commercial whaling, although the ban did not take effect until several years later. Before the conference, only a handful of governments included environmental ministries. Within a decade, more than one hundred such ministries had been created.

The UNEP was undoubtedly the single most positive and concrete product of the Stockholm conference on the international level. Despite its limited powers and resources and a variety of other handicaps, including the internal jealousies and power struggles within the UN system, UNEP proved to be of great service in its first decades to the effort to protect the global environment.

The UNEP mandate was confined chiefly to monitoring environmental developments, serving as a clearinghouse for environmental information, and acting as a catalyst and coordinator within the UN and among its member nations to carry out the agreements reached in Stockholm and to achieve other environmental objectives that transcended national boundaries. Under its first director, Maurice Strong, and his successor, the Egyptian microbiologist and government official Mostafa Kamal Tolba, who ran the agency for the better part of two decades, UNEP achieved a number of significant successes.

Despite problems, UNEP's monitoring activities made substantial contributions to achieving a clearer vision of the state of the global environment and the threats confronting it. The Earthwatch system it set up included a Global Environmental Monitoring System (GEMS), which enlists thousands of scientists around the world to assess the health of ecosystems on a continuing basis; INFOTERRA, which disseminates information on a broad spectrum of environmental issues, and the International Register of Potential Toxic Chemicals.

Peter S. Thacher of the World Resources Institute, a former deputy director of the environmental program and a veteran environmental diplomat, concluded in 1991 that, "thanks in large part to progress made since Stockholm under UNEP's 'Earthwatch' activities, a basis now exists on which to assess risks and the likelihood of harmful impacts, and arrive at cooperative measures for reducing them or taking precautions.[9] UNEP played an important role in developing scientific evidence of the threats of global warming and destruction of atmospheric ozone and in alerting the international public to those dangers.

The UN group also acted as a sponsor for a variety of multinational conferences, programs, plans and agreements covering such diverse agendas as human settlements, water resources management, transboundary movements of air pollution and hazardous waste, biological diversity, land degradation, desertification, environmental education and environmental law.[10] One of the earliest and most successful UNEP initiatives was its Regional Seas program, particularly its Mediterranean Action Plan, with which the nations on the littoral of that fabled but troubled body of water joined forces to address its many pollution problems. Because of the initiative, normally hostile states such as Israel and Syria, Egypt and Libya, and Greece and Turkey sat peacefully at the same table to try to protect their shared resource and heritage. Stjepan Keckes, head of the program, observed that a regional approach to problems enabled disparate nations to focus on problems they face in common. "Pollution is the unifying concept," he explained.[11]

It was clear from the outset, however, that UNEP would not be the answer to the planet's environmental ills. Aside from the fact that it was deliberately given a limited role, with no real operational capacity, it labored under several handicaps that made it difficult to carry out even its assigned functions. One was the fact that it had no real authority within the UN system. Another was that, to accommodate a desire by Africans to have a UN agency on their continent, UNEP made its headquarters in Nairobi, Kenya, far from the diplomatic power centers of New York and Geneva. A number of critics contended that the autocratic, excessively centralized leadership style of Mostafa Tolba, who headed UNEP for seventeen years, seriously compromised its effectiveness.[12] But an argument can also be made that, given its handicaps, UNEP would not have been able to accomplish as much as it did without his very firm leadership. Tolba's personal diplomacy certainly influenced several important environmental negotiations, including the decision at Montreal to phase out chemicals destroying the earth's protective ozone layer.[13] UNEP's most onerous problem, however, was its lack of adequate funding. It was originally supposed to operate at a level of $100 million a year. But for many years it had a budget of only $30 million, less than some of the national nonprofit environmental organizations in the United States.

But even without these handicaps UNEP, or indeed the UN system as a whole, was overmatched by the mounting complexity, scope, and seriousness of the threats to the global habitat. The UN system, divided into specific areas of responsibility and fiercely defended bureaucratic fiefdoms, was incapable of the kind of cross-sectoral approach demanded by these interwoven economic, ecological, and social dilemmas. Even more to the point, as Tolba noted, "the global problems are created and must be solved nationally within a global framework." But within national governments, he noted, environmental ministries have much less "clout" than do economic, finance, development, energy, industry, and agriculture ministries.[14] And the framework itself—the UN—could not operate effectively without "Olympian leadership" from a national government or group of governments, said Brian Urquhart, the diplomat who served for decades as under-secretary-general of the UN. In recent years that leadership has been absent, he lamented, in large measure because the United States, which had led the world in the drive toward internationalism in the post–World War II era, no longer filled that role.[15]

The erosion of America's enthusiasm for its role as world leader began with Vietnam and was accelerated by the energy shocks of the 1970s and the economic slowdowns that bedeviled much of the 1970s and 1980s. The triumph of right-wing politics with the election of Ronald Reagan in 1980 gave new impetus to the drift toward neoisolationism on issues such as the environment and development assistance. Even at home, the federal government turned aside from its commitment to environmental and social justice. In 1982, ten years after Stockholm and two years into the Reagan presidency, Russell Train, chairman of the World Wildlife fund, a former head of the EPA, former chairman of the Council on Environmental Quality, and a Republican, noted that in its environmental activism, "the United States was light years ahead of the rest of the world. Then it stopped. . . . Presidential leadership is missing."[16]

Environmentalism itself, which was considered as American as apple pie, began to be attacked from what was then the fringe of domestic politics. With the decline of Soviet power, the radical right in the United States, which exercised increasing influence within the Republican Party, turned away from its preoccupation with communism to focus on environmentalism as the new devil around which to build its membership drives and fundraising efforts. In 1990, Patrick J. Buchanan's newsletter, *From the Right*, an intemperate, aggressive, but representative voice of ultraconservative politics in the United States, carried "An Anti-Environmentalist Manifesto." It warned that in environmentalism "we face an ideology as pitiless and messianic as Marxism. . . . [E]nvironmentalism harks back to a godless, manless and mindless Garden of Eden." Human existence, it pronounced, is based on "subduing nature."[17] It seems not to have occurred to the author, who apparently was not aware of any change in scientific perspective since the sixteenth century, that nature might be subdued to the point where human welfare was imperiled.

As the right came increasingly to dominate the Republican Party, which, under President Theodore Roosevelt had originated federal efforts to safeguard the nation's land and resources, Republicans sought increasingly to roll back the progress made in environmental protection in the United States during this century.

With apostles of free-market infallibility in political power in the United

States, the United Kingdom, and several other industrialized nations, spending on the environment and governmental regulation to curb pollution and unsustainable exploitation of natural resources was cut back sharply. Environmental institutions, already the neglected children in governmental hierarchies, were starved for funds and weakened still further, particularly institutions like UNEP that are engaged in international environmental activities. President Reagan, for example, sought to block all funding for the UN environmental agency but was thwarted, to a degree, by Congress, which restored a part of the U.S. contribution.

Policies and budget allocations for national and collective security, instead of expanding to reflect new understandings about the destabilizing effects of poverty, the vulnerability of global life support systems, and the common dangers facing all of humanity, contracted to almost single-minded emphasis on military security. An orgy of spending on armaments, led by the United States during the Reagan administration, spread across the world, including—one might say, especially—in the poor countries that could least afford them. In 1984 arms imports into Africa, a continent in which hunger was and is rampant, exceeded grain exports for the first time.[18] UN secretary-general Boutros Boutros-Ghali, discussing famine and clan warfare in Somalia in the early 1990s, said: "There are more arms than food in Somalia. These arms were not fabricated by Somalis . . . they were given by the outside world to serve their own interests. Those who provide arms are partners in crime."[19]

By 1985 global spending on armaments reached the almost unimaginable sum of $1 trillion a year.[20] By 1990 such spending had tapered off a bit but was still well over $900 billion. Aside from the actual and potential harm to human life, human health, and the human environment that such arms spending represented, it also drained away money desperately needed to spur development in the poor countries of the world and to take action against the mounting ecological threat to the well-being and security of the human family.

In his book *Ultimate Security,* Norman Myers suggested some of the trade-offs that are required because of this drenching of arms makers in public money. It would take, for example, $40 billion a year, less than annual spending by the United States on the Strategic Defense Initiative during the Reagan administration, to improve the agricultural capacity of the developing world "to an extent that would allow all people to go to bed with a full stomach." To reverse the spread of deserts would cost $12 billion a year, or four days' spending on the military. A five-year global child immunization program that would prevent a million deaths a year would cost $1.5 billion, or what it cost to fight the Persian Gulf War for one day.[21]

Even as the Cold War ebbed and arms spending began to slow slightly, a huge legacy of deadly stockpiles of nuclear, chemical, and biological arms remained. Not only do these stockpiles represent a grave threat to public health and safety, their disposal will require enormous additional expenditures of public treasure. In the United States alone, it has been estimated that cleaning up contaminated nuclear weapons facilities could cost $300 billion or more, not counting the permanent storage of radioactive wastes.[22]

As the twentieth century draws to a close, it is painfully apparent that armaments do not assure collective, national, or individual security to anyone on earth. In this century alone tens of millions of people, a majority of them noncombatants, have lost their lives in increasingly industrialized, efficient, and impersonal warfare. And as Jeremy Rifkin has pointed out, "The nuclearization of the global commons has made the present generation the least secure in history."[23] Moreover, there is ample evidence that the huge expenditure of national treasure for unproductive military purposes caused the United States and the Soviet Union to lose their relative economic and geopolitical strength, as it almost inevitably has done throughout history to nations that seek to extend their power through force of arms. Yet spending to reverse environmental deterioration, poverty, and economic and social inequity, recognized in Stockholm and subsequently as real and present dangers to our security and common welfare, remains a fraction of spending on armaments in most countries of the world.

. . .

Meanwhile, the economic health and political stability of the developing countries, identified in Stockholm as essential for safeguarding the global environment, improved little if at all and in some areas, notably Africa, actually deteriorated. At the same time, the gap in wealth, consumption, and quality of life between the rich and the poor nations continued to widen.

The record is all the more tragic because many of the developing countries, collectively called the South, seemed poised for economic takeoff at the beginning of the 1970s. The process of decolonialization was nearly complete, and in most of the former colonies the scars of war and of cultural exploitation had pretty much healed. The green revolution promised to remove the scourge of famine. Booming economies in North America, western Europe, and Japan offered a somewhat more receptive market for the primary producing nations. The World Bank and other multilateral institutions, as well as many of the rich countries, expanded their development assistance. And the economies some of them, chiefly on the Pacific Rim of Asia, did manage to take flight.

But the development process began to turn sour early in the 1970s, although in many countries the effects were not fully felt until a decade later. Many, but by no means all, of the reasons for the failure were far beyond the control of governments and peoples in the South. The turnaround in the fortunes of both North and South began in the early 1970s with the abandonment of fixed currency exchange rates, followed by the "oil shocks" created by the sudden steep increase in world oil prices imposed by the Organization of Petroleum Exporting Countries (OPEC). What ensued was a massive destabilization of the global economic structure, a crisis marked by widespread inflation, declines in industrial output, demand, and investment; a sharp jump in unemployment; and rapidly deteriorating international payments balances in many countries.[24]

The rich countries were affected by these economic traumas, but the impact and suffering were greatest in the poorer nations. While the South's share of world's total production of goods and services rose from 15.9 percent in 1970 to 17.8 percent in 1987, little progress was made in improving the quality of life in the developing nations because population growth often outstripped increased productivity.[25] Between 1970 and 1990, the South's share of the world population rose from 71.6 percent to 77.2 percent (and was projected to rise to 84.1 percent by 2025).[26] This was particularly true in sub-Saharan Africa, where population growth exploded at a rate of nearly 3 percent a year in the 1970s and 1980s.[27] Per capita gross national product (GNP) in that sorely pressed area of the world fell from $490 in 1970 to $440 in 1987 after accounting for inflation.[28] (It must be acknowledged that the United States bears at least a measure of responsibility for the population crisis in the South. The United States, which had been a leader in global family planning efforts, withdrew support for the UN Population Fund during the Reagan and Bush administrations, whose policies equated population control with advocating abortion.)[29]

Prices of the commodities the developing countries exported to obtain hard currency for imports and for investment capital plummeted. The world market price for coffee, sugar, most foods and fibers, even energy dropped sharply from the early 1970s, sometimes by 50 percent or more.[30] Only a few commodities, such as timber and some metals, increased in price. With each drop in price the hardships of farmers, miners, and other workers increased, and the pressure they were forced to place on the land, the forests, and the fisheries to increase production intensified. The inability of farmers in much of the Third World to support their families off their land drove hundreds of millions of people into the cities, creating seemingly unsolvable problems of crowding, unemployment, poverty, drug addiction, crime, and disease. The rapidly growing cities of the developing countries also produced fearful pollution, which governments were unable to address because they lacked the funding for even basic control technology.

In 1968 the UN set a target of 0.7 of GNP for the rich countries to set aside for development assistance to the poorer countries. The United States, then the world's biggest aid-giver, did not agree to accept the goal. The target was met only by a couple of northern European countries, and in the 1980s the amount of aid from the industrialized nations slipped from 0.37 percent of GNP to 0.35 percent, half the UN target.[31] Because of falling commodities prices, deteriorating terms of trade, declining development aid, failed development strategies, and sometimes misdirected assistance from the World Bank and other multilateral institutions, more and more of the poor countries found themselves falling deeper in debt to the rich. By the early part of the 1980s, service on this debt was largely responsible for a net *outflow* of capital from the South to the North that approached $50 billion annually. Sir Shridath Ramphal, a former secretary-general of the British Commonwealth, likened this bizarre state of affairs "to that of a poor family in bondage to a rich money-lender, eking out an existence without hope of end."[32] The late Willy Brandt of Germany said this transfer of wealth from the poor to the rich nations was like "a blood transfusion from the sick to the healthy."[33]

As all of this suggests, after Stockholm the rich countries, generally speaking, continued to grow richer and the poor poorer. In 1970 the richest 20 percent of the peoples of the world, most of them living in the industrialized countries, controlled 70 percent of the global income, while the poorest 20 percent, almost all in the developing countries, accounted for 2.3 percent. By 1989 the rich had taken control of 82.7 percent of global income, while the poor's share had dropped to 1.4 percent.[34] The developed countries, with 24 percent of the world population, consumed a hugely disproportionate share of the world's material goods, including 61 percent of its meat, 81 percent of its paper, 92 percent of its automobiles, and 75 percent of its total energy. And of course, the rich countries account for most of the world's pollution and waste.[35]

The growing power of the transnational corporations also tended to concentrate wealth in the hands of a shrinking number of people, mostly living in the North. Just fifty of these giant conglomerates account for half of all private foreign investment around the world.[36] In the 1970s and 1980s only a few of the transnationals developed any kind of strategy for dealing with the environmental consequences of their activities, much less assumed any responsibility for the well-being of the people of the countries in which they operated.[37]

Those are not just abstract numbers. They tell of gross imbalances in the way people in different countries live; of deprivation and misery, overindulgence and extravagance; how much and what kind of food people eat; the adequacy of their shelter, their clothing, their health, and their education; how their women are treated; whether their children live or die; whether they experience happiness or pain; and the number of years they may expect to live on this earth.

Both overconsumption by the rich and underconsumption by the poor are now recognized, along with population pressures and destructive technologies, as key causes of environmental decline. The industrialized consumer nations engorge the world's resources and spew its wastes and poisons into the environment at rates that, by wide if not universal consensus, cannot be sustained indefinitely. The poor of the developing countries use up and destroy the trees and soil and water to keep themselves alive for today, but by so doing they are reducing their prospects for survival—much less prosperity—in the future.[38] "Eating the seedcorn" is the way Barbara Ward described this dilemma.[39]

To many, particularly in the South, the growing gap between rich and poor is seen as a terrible injustice inflicted by the greed of the industrial nations. An angry African expression of this perspective was given by Louisa Tappa of Cameroon, who wrote in 1991 that

> instead of development, it is my view that we are in the most vicious stages of colonization. The so-called development strategies are ways to exploit the riches of Africa. For industrialized countries, Africa is simply a vast plantation, a huge market, innumerable sales points, an immense dumping ground, a place for leisure, and a reservoir of labor and jobs. She will therefore have to produce raw materials and export them to the industrialized countries for a pittance; and she will get so much the poorer by consuming manufactured goods purchased at incredibly high prices from the very same countries.[40]

There can be little doubt that the industrialized nations of North America, Europe, and Asia, which control the lion's share of the planet's wealth and power, also bear the major responsibility for inequitable distribution of its wealth, degradation of its environment, and heightened political tensions between North and South. But the countries of the developing world are not without sin. The report of the South Commission, whose members came entirely from those developing countries and whose chairman was Julius K. Nyere, the respected former president of Tanzania, noted that "lack of democracy, corruption, and militarization, further eroded the economic and political bases for development."[41] One veteran observer, a former senior government official from an industrialized country, wrote: "So effective is this use of corrupt power in the developing world that a sum of U.S. $600 billion, or two-thirds of the total Third World debt burden, is now estimated to have been successfully diverted to private, unauthorized channels."[42]

Apologists for corruption in the South argue that it is just as endemic if not more so in the rich industrialized nations and that criticism of those practices in the developing countries is hypocritical cant and an excuse for cutting back on assistance. Look, they say, at the Ivan Boeskys and the Michael Milkens in the United States, the discredited governments of Japan and Italy, and the "you scratch my back and I'll scratch yours" relationship that exists between government and business in virtually every northern capital. The argument is true enough, but it is largely beside the point. Corruption in rich countries is reprehensible and criminal, but in the developing countries it is more than that. Corruption and the means by which it is exercised can be life-and-death matters in poor countries where people live close to the edge of survival. It can mean a prolongation and widening of economic and social injustice. It can mean the difference between nurturing and improving ecological systems and destroying them. It can mean the difference between building a nation and a descent into anarchy. One has only to see what happens when the president of Zaire arrogates much of his country's wealth to his own personal fortune; when the warlords of Somalia fight for power, using their starving nation's foreign exchange to purchase weapons; when the political and military power of any number of Central and South American countries at various times is used to protect the great landholdings and privileges of a small elite class; when the Marcoses skim the wealth of the Philippines, a country of abundant natural resources and widespread poverty, to accumulate a vast private fortune. For those and many other nations, corruption is obviously far more than an issue of morality and of crime and punishment.

. . .

By the mid-1980s not even the most sanguine observers could overlook the failures of the development process and the flagging cooperative efforts to safeguard the global commons. There were frequent, sometimes dramatic, occasionally catastrophic reminders that the hopes and promises of Stockholm were not being fulfilled. Industrial air pollution was blamed as the chief cause of the *Waldsterben,* the death of the forests, afflicting much of central Europe. A discharge of methyl isocyanate from an American-owned Union Carbide chemical plant in Bhopal, India, in 1984 killed several thousand men, women, and children and maimed, blinded, or otherwise affected as many as two hundred thousand more. Most of the dead and injured were the very poor, whose destitution was the chief reason they lived in proximity to the plant.[43] Two years later a fire at a chemical warehouse in Basel, Switzerland, resulted in the release of thirty tons of highly toxic pesticides into the Rhine River. A 120-mile stretch of the river was left "biologically dead for several years," hundreds of thousands of fish were killed, and drinking-water supplies from Germany to Holland were at risk.[44] The meltdown of the nuclear reactor at Chernobyl in the Soviet Union (now in the Ukraine) spread radioactive fallout across Europe, from Scandinavia to Italy and as far west as the British Isles. The reactor had been poorly designed and poorly maintained by an increasingly indigent Soviet government. The toll of this disaster is still not known. Some Soviet doctors have said that 3.5 million people are at risk for cancer or leukemia, but it can take many years for those diseases to develop.[45]

Catastrophes such as these focused attention on the planet's growing ecological crisis; however, the real causes were not accidents but the planned workaday economic activities of human society and the technology used to carry them out. Perhaps nowhere was the application of powerful technology to the service of misguided economic goals and ideological hubris more destructive than in Russia and its former satellite states. Many of the economic failures that led to the collapse of communism in Russia and Eastern Europe can be traced to ignorance of and indifference to the environmental consequences of their development strategies.

A poignant illustration of this failure is the fate of the Aral Sea. The Aral once covered twenty-six thousand square miles and was one of the most productive fisheries in Asia. But under economic planning initiated by Stalin, the rivers that feed the sea were diverted to provide irrigation for cotton and other crops. Over the years the lake receded and now covers only eleven thousand square miles. Photographs show the lonely sight of fishing vessels stranded on a sandy desert twenty miles or more from the water. Most of the fish are gone, and the fishermen are without work. And the salt left exposed on the dry seabed is carried by the wind to the thousands of square miles of the surrounding agricultural land, rendering them incapable of growing cotton or anything else.[46] Mikhail Kokeev, director of environmental affairs in the foreign ministry just before the collapse of the Soviet Union, explained that Stalin's successors had followed his ambition to "reconstruct nature" in the service of Socialist development. "The idea," Kokeev said, "was to build new ecological systems, new channels new rivers, everything. Of course it failed because it was not realistic."[47]

Much of eastern Europe had been turned by a brutal lunge toward industrial expansion into reeking dead zones of barely breathable air, corrosive river waters, dying trees, eroded soils, crumbling nuclear plants, unhealthy workers, diseased children, and, of course, broken economies.[48] It was small consolation that economic failures also left some areas of eastern Europe relatively pristine, whereas almost all of the rest of Europe had been touched by development in the post–World War II prosperity.

In sub-Saharan Africa, the human suffering caused by what former World Bank president Robert S. McNamara called the "vicious cycle" of explosive population growth, agricultural stagnation, and degradation of the natural resource base was even greater. In many African countries, McNamara wrote,

> rapid population growth is putting unbearable pressure on agricultural lands. In many countries, traditional farming land is already over-cultivated, and more fragile lands are being exploited. In many countries, poor families are cutting whatever wood they can lay their hands on for essential fuel. The result is ever-widening circles of bare and infertile soil around villages, ever more time and effort required simply to obtain fuel and raise enough crops to survive, and less time and energy for improving human welfare.[49]

The environmental assaults went on and on, continent by continent. In Asia the hunger of Japan and other rapidly expanding economies in Taiwan and South Korea for timber led to the stripping of forests in Malaysia, the Philippines, and other tropical areas with the eager assent of governments seeking to accumulate foreign exchange. Japan's demand for timber and the U.S. taste for imported beef, as well as the need of the landless poor for acres to till, led to wide, rapid deforestation in Brazil, Mexico, and other areas of South and Central America. The global acreage available for growing grain has been decreasing since 1950 as land is being preempted by urbanization, suburbanization, and industrialization. After peaking in the mid-1980s, the absolute number of acres devoted to grain production has declined.[50] The use of drift-nets and high technology by fishing fleets strip-mined life in the oceans to the point where per capita fish catches began declining. The teeming, seemingly inexhaustible schools of codfish, one of the great lures that attracted Europeans to North America and provided food and prosperity to their descendants for nearly five hundred years, dwindled to the point that cod fishing off the United States and Canada had to be limited and in places suspended in the early 1990s.[51] The continued expansion of automobile fleets in North America and Europe and their insatiable combustion of petroleum made clean air goals an ever receding target and inevitably led to ecologically destructive accidents such as the *Exxon Valdez* oil spill in 1989, which sullied Alaska's beautiful and productive Prince William Sound. The barbarism of Saddam Hussein of Iraq in torching the oil fields of Kuwait after losing the Gulf War once again demonstrated the vulnerability of regional ecosystems to acts of military savagery. The threat of "nuclear winter," the rapid change of climate that would follow a major nuclear exchange, raised the possibility that much of life across the entire planet would be threatened with extinction in the event of war.

The new attention paid to human assaults on the environment after Stockholm—UNEP's monitoring activities; the activities of the new national environmental agencies; the work of a growing number of academic, governmental, and industrial scientists; and the intense scrutiny of the swelling army of environmental activists around the world—alerted governments and the public to a new set of insidious but potentially ominous threats to the global environment and the life it supports.

The threat to the earth's protective ozone shield caused by chlorofluorocarbons and other industrial chemicals, the prospect of global warming as a result of emissions of carbon dioxide and other greenhouse gases, the mass elimination of species because of deforestation and other economic activity, and the pollution and toxification of every corner of the planet, including the remotest polar areas, from agricultural and industrial effluents brought home the increasingly alarming message that the entire planet and all of its inhabitants were at risk from human activity.

. . .

Two decades after Stockholm it was clear, as Peter Thacher observed, "that humans are shaping their environment at all scales—local to global—in ways that risk adverse effects to future generations."[52] A group of American physicians found in 1993 that the "threat to human health and survival" posed by the changes in the global environment is a problem "more complex by orders of magnitude than that of nuclear war, and its solution will demand much greater changes in the way people lead their lives, in developed and developing countries."[53]

The "only one world" vision of the Stockholm meeting turned out to be a chimera that quickly faded from the stage of international power politics. National governments failed to grasp the possibilities of cooperation in the face of a common enemy and quickly reverted to the old patterns of competition and zealous opposition to any hint of sharing their sovereignty. The old frontier economics of growth at any cost and unlimited exploitation of resources continued to dominate national policies and economic relations among nations.[54] The international legal system and international institutions, particularly the UN and the Bretton Woods system, were sadly incapable of meeting the challenge. Local demands on the global commons overwhelmed the shared international interest in its preservation.

Writing in the mid-1980s, Lynton Keith Caldwell found an "immense disparity" between the rhetoric of governments and international institutions and the reality of a threatening environmental future. "It is hard to avoid the conclusion that the changes in human expectations and behaviors required to achieve sustainability are greater than humanity is able to make in its present state. If mankind's world is to escape self-destruction, social changes not now foreseeable would have to occur."[55]

Starting in the 1970s and continuing through the 1980s, many nations, chiefly in the industrialized North began to address and make modest improvements in local and regional environmental problems such as air and water pollution. But global threats such as the greenhouse effect and ozone depletion were largely ignored and continued to worsen.

By the end of the 1980s, however, change, or the possibility of change, was perceptible. The ecological and economic failures of the preceding years were helping to produce around the world what Jessica Tuchman Mathews described as "a fundamentally new appreciation of the importance of the environment to the human condition" that would "gradually force us to shift our focus from an overriding concern for the welfare of our own species to that of the planet as a whole."[56] As Mark Sagoff observed, "we are now responsible for the course of evolution; in the age of biotechnology and a global economy, it would be absurd to think we could avoid this responsibility."[57]

This new appreciation arose from a number of sources: accumulating scientific evidence of the dangers, repeated confirmation of the relationship between economic and ecological health, a growing sophistication in the news media and their willingness to report on the problems, and the courage of a handful of political leaders willing to press the issues in national and international forums.

Above all, however, pressure to confront the global ecological/economic crisis came from below. It came from grass-roots organizations whose members experienced for themselves the threat to even their bare-bones livelihood because of environmental degradation and whose children sickened and died because of pollution. It came from national and international environmental organizations that refused to let governments turn their backs on the problems and that aroused a lethargic public opinion to demand that those in power take action.

Chapter 5

Fire from Below

> *Never doubt that a small group of thoughtful, committed citizens can change the world. Indeed, it's the only thing that ever has.*
>
> —Margaret Mead

In Washington, D.C., a group of well-tailored public-interest lawyers and scientists with degrees from Ivy League schools sit down with State Department officials to discuss the U.S. negotiating position on a global warming treaty.

In the hills of Uttar Pradesh in India desperately poor villagers wrap their arms around trees to preserve them from the logger's ax.

In Poland, just before the demise of the Communist state, thirteen citizens are arrested for demanding that the emission of toxic metal dust from a local foundry be halted.

In London, economists, agronomists, and planners, sitting in the well-appointed offices of a nonprofit think tank, prepare a report on sustainable agricultural practices.

In Nairobi a group of women are beaten and arrested when they stage a sit-down in a park in the heart of the city, where the government plans to build a skyscraper despite the city's extreme need for open space.

In Lithuania, Estonia, and Latvia, then still under Soviet domination, tens of thousands of citizens form a human chain along the Baltic shore to protest the gross, destructive pollution encouraged by the Moscow's reckless industrialization drives.

In São Paulo, Marco Tereno, a Yanomami Indian, holds a news conference to urge that his people be protected from settlers, loggers, and miners and be permitted to pursue their way of life, which does not destroy their Amazon Basin habitat.

In the north of Canada, Chippewa Indians campaign to mobilize public opposition to a giant hydroelectric project that will flood their hunting lands.

In Paris, "Danny the Red," a left-wing radical who helped foment the antigovernment student riots in the 1960s, is now an activist in the French Greens, who are seeking to pursue their environmental and social goals through the elective process.

In Botswana, members of the Kalahari Conservation Society fight to block diversion of the Okavango River, which supports one of the great remaining wildlife habitats in Africa, from being diverted for use by a diamond mine.

In Miami a group of women from all continents gather to form a worldwide network that would bring feminine knowledge and sensibilities to bear on the task of preserving the global environment.

In Brazil's Acre Province, Chico Mendes, leader of the local rubber tappers, fights to prevent wealthy landholders from cutting down the forest that sustains his fellow workers. He correctly anticipates that he will be murdered for his efforts.

In Zurich a group of powerful chief executives of some of the world's biggest transnational corporations meet to discuss how environmental and demographic trends will affect their businesses in the future.

In the rain forest of Colombia isolated villagers, with funding from Conservation International, an American environmental group, carve buttons and other objects from the nut of a native tree. The project not only brings in badly needed cash, it also provides a sound economic reason for keeping the loggers and developers out of their forest and leaving the trees standing.

In New Hampshire, in Japan, Germany, Britain, and Russia, local citizens organize to try to block the construction of nuclear power plants.

. . .

These are very different peoples, with different belief systems, different values, different needs, different problems, different agendas, and very different capacities for carrying out those agendas. Some are concerned with local predicaments, others with global threats. Some are engaged for ethical reasons, some out of professional or career interests, and others because their livelihood, their health, their safety, and even their survival are at stake. Not only because of language barriers but because of different cultures and widely divergent worldviews, many of them could not talk to each other, much less argue with each other.

Yet all of them are members of a broad and growing social impulse that is flowering around the world. It is a global environmental movement, yet it is more than that. Its roots lie in the wildlife conservation tradition, in the fear and anger caused by industrial pollution and resource degradation, and in the increasing public awareness that human activity is imperiling the earth's life-support systems. But those roots are also embedded in the rising demand of people around the world for economic justice, due regard for their human rights, and above all, a voice in the way their world is ordered. It is largely but not entirely a movement of people who operate outside of government joined in what are frequently called nongovernmental organizations (NGOs) or social movement organizations. Together they are sometimes called the independent sector or the civil sector. Usually, their goal is to influence government policies; often it is to take direct action to help themselves.

. . .

It is no accident that environmental movements arose first in countries with strong democratic traditions, especially the United States, Canada, and Britain. In the United States the birth of the conservation movement at the end of the nineteenth century coincided with the rise of populism and the emergence of progressive politics under Theodore Roosevelt and their insistence on the democratic allocation of the nation's resources.[1] In England, the first country to feel the full physical impact of the industrial revolution, public activism began in the middle of the last century.[2] Environmentalism became a deeply held value in other countries with effective democracies, including post–World War II Germany, the Netherlands, Scandinavia, and Switzerland.

Environmental activism in the developing countries, virtually nonexistent before Stockholm except by western European and North American wildlife organizations, took hold in many eastern European, African, Asian, and Latin American nations in the 1970s and 1980s. In part this was a legacy of Stockholm and increasing information and education about the environment and its relationship to economic development. But it also paralleled an unexpected overthrow of authoritarian and military regimes and the installment of democratic governments in many of those countries, often for the first time in their history. As Enrique Iglesias, president of the Inter American Development Bank, noted in 1989, the year the Soviet empire suddenly crumbled, "We are in a world of surprises. Man has made a dramatic reappearance on the political stage. The individual is clamoring for human rights, for physical, cultural and civil rights and demanding a voice in world affairs."[3]

Over the past two decades those voices have increasingly been raised in a demand not only for basic human rights but for environmental as well as economic justice. In 1992 a twenty-four-country public opinion survey taken by the George H. Gallup International Institute found that concern about the environment and desire to do something about it was just as strong—and on some issues stronger—among people living in the poor developing nations as among those in the rich industrialized nations. Respondents in the poor countries expressed themselves just as willing as their richer counterparts to make economic sacrifices for the environment. "The conventional wisdom is wrong about the existence of major differences in levels of environmental concern between citizens of rich and poor nations," the authors of the survey wrote and added: "Indeed, protecting one's family from environmental hazards seems to be joining the provision of food clothing and shelter as a basic human goal."[4]

It is also no accident that environmental assaults and despoliation of resources have been fiercest in areas such as Haiti, the Horn of Africa, the former Soviet Union, and China, where democratic institutions were weak or absent. As *Defending the Earth*, a 1992 report by the Human Rights Watch of the Natural Resource Defense Council, amply documents, the environment is most vulnerable where human rights are routinely violated.

The report notes, for example, that when Vaclav Havel became president of the newly democratized Czechoslovakia in 1990, in one of his first speeches he told the Czech people: "We have laid waste to our soil and the rivers and forests that our forefathers bequeathed to us, and we have the worst environment in the whole of Europe today." What Havel did not have to mention, the report said, was that "Czechoslovakia's natural environment could be poisoned with impunity because of the previous regime's near-total control of citizen organizations, the press, universities and other potential sources of oversight and criticism."[5] Similarly, in a 1991 Panos Institute report on the effects of apartheid in South Africa, the scientist Mamphela Ramphele wrote that "participatory democracy is a vital prerequisite for the upgrading of the environment, enabling people to reclaim control and to hold authorities accountable for the communities they purport to serve."[6]

But if the environmental movement can flourish only in a democracy, environmentalism and environmentalists have often been at the cutting edge of democracy movements; their demands for environmental justice have been the first popular voices to challenge authoritarian governments in a number of countries. Nowhere has this been more evident than in the former Soviet Union and its eastern European satellites. Independent citizens groups organized on such environmental issues as Moscow's plans to divert Arctic rivers were the first stirrings of a popular voice in decision making during the *glasnost* period. The gross pollution in East Germany, Hungary, Czechoslovakia, Poland, and other eastern European countries produced hundreds of "ecology clubs" that otherwise repressive governments could not directly attack.[7] The calamitous nuclear explosion at Chernobyl is believed by some to have hastened the downfall of the Soviet regime.

Ecological activists are also in the forefront of the struggle for human rights and democratic societies in a number of developing countries. The Green Belt Movement of the National Council of Women of Kenya, led by Dr. Wangari Maathai, not only enlisted women to help restore a degraded environment but encouraged them to make their own independent decisions in a country with a corrupt, quasi-authoritarian regime, which has tried unsuccessfully to ban the movement.[8] In China, Fang Lizhi, the dissident astrophysicist who was jailed for his outspoken advocacy of democratic reform, wrote that humanity will not be able to face global problems such as increasing population, environmental degradation, and atmospheric warming "as long as there are governments in the world that can hold up the slaughter in Tianamen Square as a glorious achievement, as long as there are dictators who refuse to be constrained by universal standards."[9]

There can be little doubt that the global environment movement is, as Michael McCloskey, the chairman of the Sierra Club observed, "a driving wedge for the greater democratization of life. Our movement stands for empowering citizens to share control over the conditions that shape their environment. It provides a channel of access for citizens with growing civic consciousness to participate in the procedures of government—to play a part—often for the first time—in shaping public policy."[10]

. . .

As indicated at the beginning of this chapter, the global environmental movement is extremely diverse, almost bewilderingly so. Some observers believe it cannot be called a movement at all. At one end of spectrum are big, well-funded international professional groups such as the World Conservation Union, formerly the International Union for the Conservation of Nature and Natural Resources (IUCN). The Union, headquartered in Switzerland, is a network of governmental, intergovernmental, and nongovernmental organizations from 114 countries that promotes "scientifically based action toward the sustainable use and conservation of natural resources."[11] It seeks to redirect the policies of national governments and international institutions such as the UN and the World Bank to further its conservation goals. In 1980, IUCN published its "World Conservation Strategy," which called for the preservation of genetic diversity, the maintenance of essential ecological processes, and the sustainable use of biological resources. Eleven years later, together with the UN Environment Program and the World Wide Fund for Nature, another big, powerful international conservation organization, it produced an updated strategy called "Caring for the Earth." The new strategy, reflecting the emerging recognition of the links between human welfare and the environment, emphasized both maintaining the capacity of the earth's life-support systems and improving the quality of human life.[12]

At the other end of the spectrum are the thousands upon thousands of grass-roots organizations composed of people who have banded together to improve their supplies of fresh water, to save the forests they depend on for food and fuel, to block construction of a dam that will flood their land and drive them from their homes. These groups will oppose toxic waste dumps in their neighborhoods, oppose development that threatens their land and native wildlife, and develop simple technologies such as cooking stoves and irrigation tools that will save money while easing the strain on resources. They will plant trees and open fish farms, they will petition governments to carry out environmental protection laws, and they will engage in civil disobedience when the government is an agent of environmental destruction or go to jail when an authoritarian government will not tolerate their protest.

No one really knows how many people belong to these grass-roots groups, although Alan Durning of the Worldwatch Institute believes they number in the hundreds of millions.[13] In the United States, Europe, and Japan, these groups are able to attract the attention of the news media and government officials and to influence the political process as a means of achieving their goals. In the developing nations and in countries with authoritarian governments, the members of these groups are often poor and politically neutered and must overcome great obstacles to influence government policy. Taken together, however, they must be regarded as a rising, potent political force in the world, a force that is being treated with increasing seriousness by governments and international organizations.

In between are the rapidly growing ranks of national and international environmental groups, as well as research and policy think tanks and myriad other NGOs that are seeking to place their imprint on policies and institutions that a exercise control over environment and development activities.

In the United States these include old-line conservation groups such as the Sierra Club, the National Audubon Society, and the National Wildlife Federation, which have gradually expanded their activities to embrace international conservation issues and more recently have incorporated economic development goals into their agendas. They include groups such as the Environmental Defense Fund and the Natural Resources Defense Council, which germinated in the social ferment of the 1960s and use legislation, litigation, science, education, and direct persuasion of policymakers as tools to achieve their goals. Some of the groups, such as Greenpeace and Friends of the Earth, have independent branches in other countries, industrial as well as developing, and tend to take an aggressive stance of direct confrontation to push governments in a greener direction. Most of these are membership organizations, and some of them have a membership numbering in the millions. Newer organizations, such as the Global Tomorrow Coalition, the Earth Island Institute, and the Rainforest Action Network, which focus on international and global issues, have formed in response to the recognition that environmental problems can be planetary in scale. Among the more prominent of the proliferating European groups were Robin Wood and the Stiftung für Naturschutz. A number of research and policy groups, including the World Resources Institute and the Worldwatch Institute, focus chiefly on problems of global environment and development, and their data are much welcomed and used by governments around the world. All of these organizations are staffed by well-educated professionals—lawyers, scientists, writers, administrators—who generally come from a middle-class background and are relatively well paid. Several of the organizations, including the World Wildlife Fund and Conservation International, have workers doing research in the field and working with local communities in Third World countries, particularly tropical countries.

In Europe, 120 national environmental organizations with a combined membership of 20 million in the twelve nations of the European Union formed a joint "European Environment Bureau" to coordinate their activities. The International Institute for Environment and Development in London, created by Barbara Ward, produces influential reports on those issues.[14] Highly professional environmental groups are active throughout Latin America, particularly in Mexico and Brazil, where private groups such as the Funatura coalition are influential on policy matters. Coalitions of nongovernmental groups working on environmental issues also coalesced in a number of Asian and African countries, including the Philippines, Indonesia, and Kenya.

Increasingly, organizations and movements originally formed to confront other problems began in the 1980s to seek a role in environmental policies. Among the most outspoken of these were women's organizations. These groups claim a legitimate voice on ecological issues for a variety of reasons. One is the conviction held by many feminists that, unlike men, who assume a stance of patriarchal domination over nature, women are allied with nature and more nurturing toward it.[15] The late Petra Kelly, a founder of the German Greens, once asserted that "to me feminism is ecology and ecology is feminism. It's a holistic way of looking at things."[16] Another arises out of the undeniable fact that as tillers of the soil and gatherers of wood and haulers of water, women in most parts of the world have the most direct contact with the land and its resources. They are on the front lines of the fight against pollution and other environmental threats to their children and their homes. Women see themselves and their children as the chief victims of environmental degradation, but they also see themselves as the first line of environmental defense. There is also strong evidence that economic and education bias against women is a the major underlying factor beneath excessive population growth and poverty in developing countries, which, in turn, are major causes of ecological decline.[17]

The women's perspective on these issues was put forcefully by Margarita Arias, a presidential candidate in Costa Rica in 1992: "No-one speaks out for the protection of the environment with greater moral authority than women. Only those who have fought for the right to protect their own bodies from abuse can truly understand the rape and plunder of our forests, rivers and soils."[18]

Like women, organizations representing the planet's indigenous peoples claim the right to participate in decisions affecting the environment, particularly the largely unspoiled ecological systems that are their increasingly threatened homelands. There are an estimated 4,500 to 5,000 of these indigenous cultures, numbering as many as 600 million people, surviving in the world today, from the Inuits of the Arctic to the aborigines of Australia. Their homelands—rain forests, deserts, tundras, savannahs—are under extreme pressure from a rapidly expanding global economy that places in jeopardy the very existence of their cultures and sometimes even their lives. These cultures possess intimate knowledge of the uses of the plants and animals, waters, and soils of their habitat and of how to care for them, knowledge whose loss would be a tragedy for all of humankind. Accordingly, their representatives actively press for national and international environmental policies that would serve to protect their native lands. More and more they are being listened to, although not always heeded. They are increasingly vocal participants in international environmental conferences.

In recent years, religious groups from virtually all denominations have been turning their attention to the evils inflicted on the earth by human activity. An important turning point in the way theologians and the clergy perceive their responsibilities toward nature was the 1967 publication of a short essay titled "The Historical Roots of Our Ecological Crisis," in *Science* magazine, by Lynn White Jr., a historian at the University of California in Los Angeles. White contended that Christianity and, more broadly, the Judeo-Christian tradition, which established a dualism of humans and nature and held it to be God's will that nature be exploited for man's benefit, bore a "a huge burden of guilt" for the destruction of the environment by modern science and technology.[19]

White's article stimulated intense introspection in theological circles, which led, in time, to the entry of many religious groups into the environmental lists. For example, at an ecumenical meeting, "The Churches' Role in Protecting the Earth's Atmosphere," convened in Gstaad, Switzerland, in 1991, Protestant and Catholic clerics agreed that as "a faithful response to God's will" they would press for national and international commitments to reduce greenhouse gases. The conference of the clerics stated: "We believe that God calls us, our Churches, and the nations of the industrialized world to repentance for the damage already done to God's creation. . . . But we believe, also, that the Creator-Redeemer acts to give this world a future and a hope. . . . Believing in the possibility of a 'new creation' in Jesus Christ, we hear the call to a new engagement with God in establishing Justice, peace and wholeness for earth and people."

The international business community was an even more unlikely recruit into the ranks of nongovernmental environmental activists; many environmentalists, in fact, vehemently denied that businesses belonged in that category. Business and industry at the local, national, and transnational levels have been in the forefront of opposition to environmental laws and regulations that place limits on their right to consume resources and to discharge and dispose of their effluents or that might add to the costs of production. They generally fight diligently against any rules that limit their freedom of action within and across national borders.

In recent years, however, the rhetoric and often the actions of many corporations and business groups are growing perceptibly greener. I will explore this phenomenon in a later chapter on environmental economics. But here it should be acknowledged that at least a small part of the business community has thrown its support behind a realistic appraisal of and responses to the impact of its activities on the environment at all levels.

In 1990, for example, the International Chamber of Commerce, an important voice of global business and industry, adopted a "Business Charter for Sustainable Development." The basic aims of the charter are threefold: to prod businesses to commit themselves to continued improvement in their own environmental performance, to provide "common guidance on environmental management" and to aid businesses in following that guidance, and to reduce the pressures on governments to "overlegislate" by demonstrating that businesses take their responsibilities seriously.[20]

Quite clearly, some of this business response was self-serving, such as the hope of warding off national and international legislation. Just as clearly, much of it was an effort to deflect growing criticism of business as the chief instrument of resource depletion and environmental destruction. The business community, particularly the great transnational corporations, wield enormous power with comparatively little restraint. All too frequently, that power is used to pollute and degrade the environment for the sake of ever bigger and quicker profits. Industry is reluctant to yield or share this power, just as nation-states are reluctant to yield or share their sovereignty to achieve environmental goals.

But easy as it is to be cynical about the greening of industry, there is reason to believe that a growing number of corporations around the world are beginning to understand that doing the right environmental thing is in their own interest if only over the long run. A publication of the International Chamber of Commerce noted that "classical economists got away with a production function that included only land, labor and capital. . . . Today we are learning to place a proper value on our environmental capital."[21]

While trade unions have long been wary of environmentalists and environmentalism as potential threats to economic growth and jobs, there is a growing appreciation in labor circles that long-run prosperity and security for workers depends on a secure, productive environment. There is also rising anger among workers as they increasingly recognize that they are the first and often the chief victims of industrial pollution.

While much of the labor movement remains suspicious of environmentalism, it would appear that part of it at least is joining the ranks of the global environmental movement, recognizing that participation is in its own interest.

At a meeting in Tokyo in December 1990, the executive board of the International Council of Trade passed a resolution stressing "the important role that trade unions should play in the protection of the environment and appeals to governments to make 'harmonious integration of environmental and employment policies and to invest in environmental protection and technology and appropriate training for workers.' "[22]

An extremely informal but vital component of the worldwide environmental movement are "epistemic communities," or "old boys' networks" of people with similar professional interests and expertise.[23] These networks of scientists, economists, wildlife managers, air and water pollution control officers, even diplomats, are an important means of developing and exchanging information on the issues, keeping attention focused on problems—even when official governmental interest is elsewhere—and reaching consensus on what should be recommended to governments and international agencies.

. . .

Perhaps the most enigmatic components of the worldwide environmental movement are the Green political parties. Based on philosophies that reject the political, economic, social, and ethical systems embraced by the industrialized countries, and often advocating a radical restructuring of those systems, the Greens nevertheless seek to work from within. They have pursued, with some success, access to power through the electoral process; they are NGOs in government. Political scientist Horst Mewes observed that "not only do the Greens confront the momentous challenge of 'creating a new life on the foundations of ecology' out of present industrial capitalism; they are also determined to defy all odds and create a new historical kind of relationship between spontaneously formed social protest movements and an organized party."[24]

The Greens germinated in the fertile soil of an alienated post–World War II generation of young people, particularly those in western Europe who were disenchanted with the single-minded materialism of their parents. It has tap roots in the Vietnam era peace movement, the antinuclear protests, ecofeminism, Gaian principles of a holistic earth, the ethical insistence on the rights of nature of the deep ecologists, social ecology, and the E. F. Schumacher *Small Is Beautiful* approach to economics and technology, the successes of environmental activists in the United States, and the inchoate but widely shared sense of dread that late-twentieth-century civilization was on a collision course with ecological catastrophe. The Club of Rome's *Limits to Growth* and the Carter administration's *Global 2000 Report* were major documents in the movement's intellectual canon.[25]

Local Green groups first contested elections in Switzerland in 1971 and in Sweden in 1972. A regional Green party was formed in the Australian state of Tasmania in 1972, and the first national Green party, called Values was organized in New Zealand that same year. Britain's Greens, first named People and then the Ecology Party, was formed in 1973. Greens also organized in France, Belgium, the Netherlands, Scandinavia, and a number of other countries.[26] But it was the German Greens, *Die Grünen,* that made the most resounding bang in environmental politics in Europe and around the world.

Organized in 1979, by 1983 it was able to win 5.7 percent of the popular votes and twenty-seven seats in the Bundestag, the West German parliament. It earlier won electoral victories in several of the West German *Länder*, or states.[27] These successes sent political shock waves across Europe. The Greens were something new on the political stage. Their platform was based on four "pillars": ecology, social responsibility, grass-roots democracy, and nonviolence.[28] They claimed not to be classifiable under any of the old ideological labels. One of their slogans was "we are neither left nor right, we are out in front." In fact, they did include a Marxist faction and were regarded by many in Germany as a leftist movement. But some observers also saw in the Greens an echo of the pagan nature cult that was cultivated by the Nazi mythmakers of the Third Reich.[29]

Die Grünen soon split into two antagonistic factions: the "fundamentalists," who insisted that the goal must be a transformation of society, and the "realists," who believed that their program could be carried out by cooperating with other political parties to achieve reform. One result of the split is that the Greens never managed to construct a clear agenda. Their phenomenal growth of support slowed and by 1990 had fallen below the 5 percent of the vote level required to qualify for seats in the Bundestag. When Petra Kelly and her companion and fellow party founder Gerd Bastien were found shot to death in 1992—perhaps in a double suicide or a murder-suicide—it was a heavy psychological blow. Although they have won some victories in recent *Länder* elections, the concrete accomplishments of the German Greens have not begun to approach the predictions made in the first flush of their successes.

But there can be no mistaking the impact they have had on European politics. By 1989, Green parties in six nations had won a total of thirty-nine seats in the European Parliament. The Greens made strong electoral showings in Britain and France, and green politics toppled a government coalition in the Netherlands.[30] In that year a reporter for the *Los Angeles Times* discerned "an unprecedented tide of environmental concern sweeping Western Europe."[31] By 1991, Greens held parliamentary seats in Bulgaria, Romania, and Slovakia and in the former Soviet republics of Latvia, Estonia, and Lithuania and were organizing in Byelorussia, Georgia, and the Ukraine.[32]

By 1993 the fortunes of the Greens had begun to ebb. Expected to win as many as thirty seats in the French parliament, they were instead shut out. Membership in the British Ecology Party began to decline. But in part at least, the decline was a result of the Greens' successes. Both alarmed and enlightened by the popular appeal of the ecology movement, the entrenched parties in Europe began to take on a green protective coloring. Western European governments looked more carefully at the environmental price that had been paid for their postwar prosperity and began to spend some of their wealth to clean up the mess. The European Community, which had focused on economic and political harmonization, finally turned seriously to issues of regional environmental cooperation. "There's only one issue that can unite us now, and that's the environment," Gianni De Michelis, then Italy's deputy prime minister, proclaimed in 1989.[33] With their rhetoric and at least some part of their agenda coopted by the political establishment in Europe, the Greens saw their popular support dwindle.

Green parties have had minimal impact on politics in the United States, at least so far. The only time anything resembling a Green party contested a national election in the United States was in 1980, when Barry Commoner ran for president—with LaDonna Harris, of Native American heritage, as his running mate—on the Citizens' Party ticket. They received a tiny fraction of the popular vote. In recent years Green candidates have been contesting congressional, state, and local elections. In the November 1992 elections, some eighty-one Green candidates ran for office, and eleven of them, nine in California, won.[34] Before then, perhaps forty Greens had been elected to office, most at the city council level.[35]

There are a number of reasons for the Greens' lack of progress in the United States. For one thing, American politics has never been friendly to third-party candidates, particularly those who, rightly or wrongly, are viewed as espousing radical ideas. For another, the U.S. Greens have been badly splintered. While they have made serious efforts at coordination through "committees of correspondence" and the formation of coalitions and while eventual unity is still a possibility, the movement remains ideologically fractured among deep ecologists, social ecologists, ecofeminists, Native Americans, economic collectivists, and a rich assortment of other "ists." Perhaps a more significant explanation for the slow progress of U.S. Greens is that the success of mainstream environmental groups in this country in willing legislation and legal battles and achieving other environmental reforms has persuaded the electorate that the political system is responding to the environmental crisis.

In 1984, Jonathan Porritt, a leading spokesman for the British Ecology Party, wrote that "one still cannot point to any coherent green movements in the United States, let alone to any viable strategy for the long-term development of green politics."[36] A decade later the same assessment was still valid.

. . .

Despite their disunity and their inability to take political power, the Greens are a key element in the worldwide environmental movement. And that movement, as a whole, is an expanding and increasingly influential player on the global political stage. Much as governments and international institutions would like to ignore environmental problems, the environmentalists insistently keep bringing up such impolite subjects as pollution, resource depletion, the loss of biological diversity, the deterioration of global life-support systems, excessive population growth, and the links between poverty, human rights, and environmental decline.

As Caroline Thomas of Britain's Royal Institute for International Affairs noted, the environmental groups and other NGOs now play a central role in international environmental affairs and are "crucial in getting ecological issues on the international political and economic agenda."[37]

These activists influence policy in many ways. One of their key roles is to bring emerging scientific information about the environment to the attention of the public and policymakers, often through the media. For example, when governments were largely ignoring the mounting evidence of global warming in the late 1970s and 1980s, enviromentalists such as Rafe Pomerance, president of Friends of the Earth and later a senior associate of the World Resources Institute, pressed the findings of atmospheric scientists on journalists and members of the U.S. Congress and finally forced a reluctant administration to pay heed.

A number of groups conduct their own field research, particularly those such as Conservation International and the World Wildlife Fund (WWF) that are involved in ecosystem management issues. WWF data on the destruction of African elephants were a major factor in the international agreement to ban ivory trading. Data on environmental, demographic, and economic trends accumulated by groups such as the Worldwatch Institute and World Resources Institute are widely used by governments, particularly those with little research capacity, in policy-making. A report prepared by the World Resources Institute on global trends in soil erosion, for example, was cited by a number of delegations in preparatory meetings for the 1992 Earth Summit. Through the publication of books and papers and through news conferences and news releases, the NGOs play a significant role in educating the media and the public about their issues.

Environmental activists also organize citizen action, such as a 1988 boycott of fish from Iceland to protest that country's refusal to comply with the whaling moratorium.[38] They monitor the environmental performance of governments and international organizations such as the World Bank and the Food and Agricultural Organization and put pressure on them to adopt ecologically responsible policies. Sustained lobbying by environmental groups, for example, compelled the World Bank to make substantial changes in the way it lent money to developing nations. Several groups have helped arrange and even finance "debt-for-nature" swaps, through which part of the national debt a

Third World country owes to a northern lender is forgiven in exchange for an agreement to preserve land and biological diversity. The NGOs, including the business, women's, and indigenous peoples' groups, also seek to influence the outcome of international conferences on the environment and on other issues, such as trade, affected by environmental considerations. They also track compliance with international agreements such as the Convention on International Trade in Endangered Species of Flora and Fauna (CITES) and agitate for their enforcement.

Some environmental organizations try to protect the environment through direct action. Greenpeace members follow the nonviolent Quaker tradition of placing their bodies "in harm's way," for example, interposing themselves in small boats between whales and whaling ships. Sea Shepherd, a militant offshoot of Greenpeace that eschews nonviolence, tries to destroy whaling vessels without, its members insist, harming their crews. Members of Earth First! in the United States, emulating the fictional escapades of Edward Abbey's "The Monkeywrench Gang," engaged in ecosabotage against logging, mining, and development projects, at least until some of its members were arrested for planning to blow up a federal communications installation.

Other groups, particularly in developing countries, plan and carry out strategies that will help meet basic economic needs in ways that preserve the environment. For example, Kengo, the Kenya Energy and Environmental Organization, developed and distributes to poor villagers a wood cook stove that uses very little firewood, thus easing the arduous household task of fuel gathering while helping reduce the pressures on Kenya's woodlands. As a report by the UN Development Program noted, "NGOs are widely considered to be far better than governments or international agencies at mobilizing the poor for development; they are more flexible, less constrained by rules and regulations, and more inclined to directly approach the most disadvantaged in society."[39]

. . .

Given the astonishing diversity in the philosophies, politics, goals, strategies, methods, membership, and resources of environmental groups and other NGOs engaged in issues of environment and development, not to mention their geographic dispersion, it would take a great leap of faith to speak of a coherent global movement. Indeed, one scholar employs chaos theory to explain environmentalism, noting that the theory enables meaning to be drawn from dynamic, "seemingly unrelated political, economic and environmental events."[40] And in fact, the environmentalists and other NGOs are fragmented along many fault lines. Even within the industrialized nations, as we have seen among the German Greens, there are deep divisions over policies and tactics. But for many years the widest chasm was between the NGOs of North and South.

Unsurprisingly, perhaps, the differences between environmental groups from the industrialized countries and the developing countries mirror the differences between the countries themselves. Until fairly recently the northern NGOs were chiefly interested in actions to protect ecosystems such as tropical forests, to defend whales and other endangered species, and to reverse deterioration of the atmosphere, the land, and the oceans. To their southern counterparts, however, fighting poverty and achieving economic growth and an improved standard of living were the highest priorities. The demands of the northern environmentalists were viewed as, at best, indifference to the plight of poor and, at worst, as complicity in the exploitation of the developing countries for the sake of continuing the high consumption patterns of the rich industrialized nations. Southern environmental organizations accused groups from the North of attempting to impose their own agendas.

This perspective was voiced in 1992 by Hira Jhamtani, an Indonesian environmentalist, who charged that "there is an imperialistic attitude between First and Third World NGOs." U.S. environmental groups, he contended, are more concerned with "projects and campaigns than with the actual needs of Third World NGOs and communities," for whom conservation issues can be "a matter of life and death." Northern environmentalists, Jhamtani wrote, must stop supporting projects "that maintain the status quo of oppression."[41]

In their book *Global Dreams,* Richard J. Barnet and John Cavanagh lament that nation-states are disintegrating because they are no longer able to make good on Rousseau's social contract. Casting about for another center that will hold civilization together, they find that "the global environmental movement is theoretically based on a powerful unifying theme" but then add that "as a practical matter it is badly divided."[42]

. . .

Over the past few years, however, it has become apparent that even if diversity rather than unity is hallmark of the global environmental movement, substantially increased cohesion and coordination are within the realm of the possible. Several trends point in this direction. One is the spread of coalitions among the environmentalists and other NGOs, not only within countries or among northern and southern organizations but across all the fissures that divide the movement. The Environment Liaison Centre International (ELCI) in Nairobi, formed in the wake of the Stockholm conference to coordinate NGO contacts with the UN Environment Program, has experienced dramatic growth and by 1992 had a constituency of some 7,500 organizations operating in one hundred countries.[43] Old-line elitist conservation groups are now sitting down with representatives of citizens' and grass-roots organizations to plan common strategies and negotiating positions.

Most significant, the often antagonistic components of the global movement have, in recent years, started to listen to each other and even, on occasion, to hear each other. Much of this enhanced level of communication took place during preparations for the Earth Summit, which will be discussed in a later chapter. But even earlier, the policies of many groups had been changing to reflect a more expansive worldview. Environmentalists in North America and Europe, whose interest in developing countries tended to focus exclusively on issues such as destruction of the rain forest or rising levels of fossil fuel use, began to take on a broader social perspective. The Northern Alliance, an ad hoc coalition of NGOs, adopted an agenda that gave high priority to "new economics and fair trade relations between North and South, East and West," as well as to solutions to the debt crisis and the prohibition on the export of socially and environmentally harmful products and technologies, including hazardous wastes, to developing nations. The alliance also called for a change in consumption patterns in the North that are now supported by "reckless exploitation of the earth."[44]

For their part, activists in the developing countries, while continuing to insist that their problems originated in the North, began to acknowledge that they had as great a stake in addressing the ecological crisis as the rich industrialized nations had. S. M. Mohamed Idris of Malaysia, coordinator of the Third World Network, an alliance of southern NGOs, asserted that protecting the environment is not a luxury for the rich because "the destruction of the environment is going on perhaps even faster in the Third World and this ecological destruction is emerging as a major cause of poverty itself." He went on to say that environment and development issues are "parts of the same problem, the problem of the wrong model of development in both the North and the South."[45]

. . .

In December 1991 hundreds of delegates of NGOs from around the globe gathered in Paris to plan a common strategy to take to the Earth Summit in Rio. It was not easy. The meeting was paid for by the French government, which also financed the trip for activists from the developing countries who could not otherwise afford it. Many of them were attending their first international conference. For much of the time the gathering sounded like a tenants meeting on the Tower of Babel, with delegates shouting their slogans and proclaiming the primacy of their causes into the unhearing ears of the other delegates. By the end of the four-day meeting, however, a consensus on principles emerged, along with a sense of common purpose. It was stated in a draft "Citizens Action Plan for the 1990s," which began with the assertion that "a powerful common vision is emerging all over. . . . A planetary cultural ecology is being born, embodied in a full range of ethical principles."[46]

Unfortunately, the specifics of the document resembled nothing so much as a blueprint for Utopia, including plans for a draconian overhaul of the global economic system but offering little of practical relevance for the Rio conference, then just six months away.

. . .

Environmentalists are seen by some observers as the advance guard of an new postindustrial society. Sociologist Lester W. Milbrath, for example, calls them the potential "vanguard of a new society" that will replace the current "dominant social paradigm" of unbridled industrial and economic expansion and left-right political competition.[47] Al Gore, as we have seen, believes environmentalism is the foundation of a "new central organizing principle for our civilization." Certainly environmental activists and allied NGOs are listened to with increasing attention by many governments and international councils. Lynton Caldwell contends that "the importance of the NGOs in international environmental policy-making can hardly be overemphasized."[48]

The ability of the nongovernmental actors to play a decisive role in policy-making, however, remains sharply circumscribed. In many countries their voices are unheeded, and in some, particularly in the Third World, they are persecuted for challenging authoritarian governments. Their participation in councils of state are made possible only by the sufferance of governments. Their favored tactic of using the courts to achieve their goals is of relatively little utility in the international arena, where the body of environmental law is still thin and the judicial system can exercise only limited authority. They are frequently checked by powerful opposing economic forces. Their arguments are often weakened and their successes limited by the Chicken Little syndrome—too frequently, too shrilly, and much too inaccurately warning that the Apocalypse is near. Nor can the NGOs invoke the mandate of the plebiscite when pressing their positions; indeed, on many occasions they may not represent the popular will.

What is undeniable, however, is that environmentalists, far more than any other nongovernmental community except, perhaps, human rights activists, have forced their way into the previously closed rooms of international diplomacy. Even if their policies are not adopted, they are *there,* placing their position papers on the table and speaking out, not just in the corridors but in the once sacrosanct plenary halls and in the small, out-of-the-way chambers where deals are hammered out in secret meetings.

While they have not yet created a new society, it may not be too far a reach to conclude that the civil sector activists may be the opening wedge of a new, more open system of international governance. Over the past quarter of a century they have eroded, if not broken, the monopoly of governments and their satraps in the international agencies in the process of reaching decisions affecting the relationships and joint activities of nation-states.

Geopolitics, of course, is subject to sudden shifts caused by unexpected events, and trends can evaporate in response to crisis. But thanks in part to the environmentalists and fellow NGOs, it is possible to see progress toward greater democracy in the practice of international politics.

Chapter 6

(Political) Scientists

> *. . . science without conscience is but the ruin of the soul.*
> —François Rabelais, *Gargantua and Pantagruel*

Testifying in a packed Capitol Hill hearing room on a sweltering early summer day in 1988, James E. Hansen, director of NASA's Goddard Institute for Space Studies, took what was for a scientist an unusual and courageous personal risk. It was his view, he told the Senate Energy Committee, that the warming global temperature trend of recent years was probably a result of human alteration of the atmosphere. Speaking to reporters after the hearing, he was even more blunt: "It is time to stop waffling so much and say that the evidence is pretty strong that the greenhouse effect is here," he asserted.[1]

Global warming from human activity had been a matter of scientific speculation for many years. The Swedish chemist Svante Arrhenius theorized as early as 1896 that atmospheric accumulations of carbon dioxide from the burning of coal and other fossil fuels would cause the earth's temperature to rise by four to six degrees Celsius (7 to 10 degrees Fahrenheit).[2] In recent decades this "greenhouse effect," as it has come to be called, has been the object of increasingly intense scientific scrutiny. Readings taken by C. David Keeling and his colleagues at the Mauna Loa Observatory in Hawaii since 1958 demonstrated conclusively that there has indeed been a continuous, fairly steep rise in the atmospheric loading of carbon dioxide. Other scientists determined that not only carbon dioxide but gases such as methane, nitrous and oxide, and the ubiquitous industrial chemicals, chlorofluorocarbons (CFCs) could contribute to global climate change. In 1983 reports by the U.S. Environmental Protection Agency (EPA) and the National Academy of Sciences both stated that the earth would warm substantially as a result of the accumulation of the greenhouse gases. Members of the international scientific community, led by Bert Bolin of Sweden's International Meteorological Institute, agreed at a meeting in Villach, Austria, in 1985 to recommend actions by governments, including a treaty, to limit the effects of global warming.[3]

But governments did not act, neither in Washington nor elsewhere. The threat of rising temperatures stretching out over decades did not seem immediate to policymakers, particularly when scientists could not tell them what the specific effects would be in specific places. Moreover, there was by no means unanimity in the scientific community that global warming would be serious or even that it would take place at all. A number of scientists—some, but not all, funded by industry or supported by right-wing or libertarian institutions—sought to discredit the greenhouse theory and those who espoused it. And even some scientists who embraced the global warming theory stopped short of recommending that potentially costly steps be taken to counter the uncertain consequences of a changing climate. Global warming simply was not a compelling issue to the media and the public and therefore not to policymakers.

That changed dramatically in 1988. Hansen's testimony came in the middle of one of the most severe heat waves and droughts in living memory, giving the public a taste of what it might be like to live in a greenhouse world. City dwellers in much of the country gasped as temperatures climbed into the nineties and hundreds and stayed there, and smog blanketed the air with an evil, lingering miasma. Farmers in the Midwest watched their newly planted crops wither as the rains failed day after day, week after week. As river levels dropped, Mississippi bargemen sat idle on their craft, grounded by falling water. Later that summer wildfires consumed much of the parched lodge-pole pine forest of Yellowstone National Park. Other parts of the world also experienced extreme weather phenomena; in Bangladesh a ferocious tropical storm claimed the lives of thousands of mostly impoverished farming and fishing families living on the low-lying Ganges flood plain.

Hansen did not say that the climate conditions prevailing that summer were caused by the greenhouse effect, although some of his critics falsely maintained that he had done so. In fact, he told the senators quite the opposite: a specific heat wave and drought could not be blamed on the greenhouse effect, he explained, even though global warming increased the probability of such conditions. But the national media, preoccupied with the weather, seized upon Hansen's testimony—sometimes accurately, sometimes not—to give belated prominence to the policy implications of climate change. To Stephen Schneider, another atmospheric scientist who had long worked on the greenhouse problem, the summer of 1988 was crucial as a media event raising "climate consciousness" among the citizenry.[4] To the public, enduring intense heat and discomfort, global warming became a very corporeal reality. Policymakers suddenly had a hot issue, so to speak, on their hands. In part for the wrong reason, perhaps, but none too soon, the portentous issue of global climate change moved onto political and diplomatic agendas.

Hansen, a quiet, hardworking, and highly reputed scientist, developed his conclusions on climate change from a complex global circulation model that incorporated temperature readings taken over years at hundreds of locations around the world. He knew he would be criticized for going public with his conclusions; the mores of the scientific community strongly encourage its members to restrict publicizing their findings to peer-reviewed professional journals. But he did not anticipate the magnitude of the censure that would fall around his head.[5]

Much of the criticism was ideological in nature, coming from those who felt that Hansen's views could lead to government action to regulate the use of fossil fuels and who opposed government interference in the marketplace in principle. In a book published by the Cato Institute, a libertarian policy group, Pat Michaels, a climatologist at the University of Virginia, assailed Hansen for abrogating "the normal review process" and cited anonymous scientific sources to support his attack.[6] But the attacks were not just ideological; many came from colleagues who felt that a scientist should not go out on a limb in public with conclusions based on a less than completely reliable climate model. As Schneider noted, "Natural scientific skepticism, aversion to public discussion of still uncertain science, misunderstanding of the policy-making process and the media, and yes, plain jealousy" all contributed to the critical scrutiny Hansen received from his fellow scientists.[7]

Several years later, Hansen said that although his career had been affected, he had no regrets about taking a public stand on global warming. He did it, he said, not because he was on a "crusade" but because he thought it was important to call the attention of policymakers to the scientific evidence on global warming. Scientists, particularly those such as himself who are funded by taxpayer dollars, have a responsibility to speak out, he said, and added: "Publishing your findings in journals and letting it go at that is not very effective. If you don't speak out in public, it is ignored."[8]

Hansen's views were not ignored. Scientists at a meeting on climate change, a year later in Amherst, Massachusetts, almost universally condemned his "unscientific" testimony to the senate panel. "But," noted science journalist Richard A. Kerr in *Research News,* "there's an irony: had it not been for Hansen and his fame, few in public office and certainly not the public itself, would have paid much attention to a problem that everyone at Amherst agrees threatens social and economic disruption around the world. After all, experts had been hemming and hawing for a decade on the likely magnitude of the problem, and hardly anyone had listened."[9]

. . .

It is probably fair to say that, unlike Hansen, many, if not most, scientists today continue to see their professional roles as distinct from their roles as citizens. Their loyalty and their duty are to their disciplines and their research, not to the public or creation of public policy. The very language of science is frequently unintelligible to the nonscientist, helping to create C. P. Snow's mutually uncomprehending two cultures. The scientific world has its own language, methods, status system, and inbred peer group community. The culture of the humanities is built on traditions, values, symbols, and metaphors that are alien to many in the scientific community. The result, Snow found, is that the two cultures stare at each other in bewilderment across a widening chasm. Increasingly, even scientists cannot understand the arcane vocabulary of other scientists in different specialties. And as the great microbiologist and humanist René Dubos pointed out, "some of the richest human values have remained outside the fold of orthodox science."[10] In fact, science and the technologies derived from it "now often function as forces independent of human goals."[11]

Despite its growing alienation from everyday human understanding—or perhaps because of it—science and its works are accorded by many the blind faith offered to deities in earlier ages. After all, the modern world was created by science and its often unruly utilitarian offspring, technology. Indeed, the very notion of progress evolved from the scientific optimism of the Enlightenment and replaced the prevailing views of the human condition as essentially static and predetermined. As noted earlier, science was long viewed as humanity's great weapon in the struggle for survival and security against an adversarial natural world. It made possible the great expansion of human population in the past two hundred years by opening a cornucopia of agricultural and industrial production and by achieving great strides in medicine and sanitation that protected and prolonged human life.

It is understandable, therefore, that some scientists, observing a civilization built on their hard-earned knowledge, fall victim to hubris in the conviction that they are the sole repositories of immutable truth, accountable only to the integrity of their own work and to others in the priesthood of their specialized disciplines. The scientific method, they have convinced themselves, is free of any extraneous values. Any misuse of science is a result of flaws in society at large, not in their own rarefied realm.

. . .

That science and its applications can have ruinous effects on the environment and pose grave dangers to humans has, however, long been apparent to those who cared to see. More than a century ago the English writer Thomas Love Peacock complained: "They have poisoned the Thames and killed the fish in the river. A little further development of the same wisdom and science will complete the poisoning of the air and kill the dwellers on the banks. . . . I almost think it is the destiny of science to eliminate the human race."[12]

But as Stephen F. Mason remarked in his *History of the Sciences,* "The long-term consequences of the application of science were not widely appreciated before the beginning of the present century. James Watt hardly could have foreseen the urban congestion that arose from the adoption of his steam engine."[13] Thomas Midgely was one of the most honored American scientists in the first half of the twentieth century for inventing leaded gasoline and CFCs which later turned out to be two of the most ubiquitous threats to human health and the global environment.[14]

By the latter half of this century it had become increasingly apparent that science and technology are two-edged swords. Their application has had many unintended and, until recently, largely unappreciated effects, many of them harmful to human beings and the environment that sustains us—first locally, then regionally, and now globally.[15] The wizardry of science is making clever gadgets, harnessing vast reserves of energy, and manipulating the chemical, physical, and biological building blocks of creation are far outpacing basic scientific knowledge of how the world works—and of what could make it stop working. A wide gulf between the physical and social sciences has barred any systematic study of how humans and their institutions use and misuse science. By midcentury, René Dubos observed "a breakdown in the system of relationships between human nature and the scientific creations of man."[16] Some observers have also found in science and technology the seeds of the social erosion afflicting many contemporary societies. In her book *When Technology Wounds,* Chellis Glendenning lamented that "with the invention of the telephone, television, missiles, nuclear weapons, supercomputers, fiber optics, and superconductivity, the social system we inhabit has repeatedly favored technologies that usher us further and further away from the communal, nature-bound roots that for millennia honored life and interrelationship in human culture."[17]

Silvio O. Funtowicz and Jerome R. Ravetz, students of the impact of modern science and technology, noted that "few still doubt that our modern technological culture has reached a turning point and that it must change drastically if we are to manage our environmental problems." They added, however, that "it may not yet be as widely appreciated that science, hitherto the mainspring of technological progress, must also change." Because science is now widely perceived as a cause of evil as well as good, it is "losing its ideological function as the unique bearer of the True and therefore the Good."[18]

. . .

The relationship between science and contemporary environmentalism is complex and occasionally paradoxical. Foes of the environmental movement charge that it is antiscientific or, as the British scholar Anna Bramwell put it, is seeking a "return to the primitive."[19] Some critics, usually those on the far right of the political spectrum, call environmentalism an antiscientific, left-wing conspiracy or a new religion of the apocalypse (at the same time predicting that environmental regulation will produce an economic apocalypse).[20]

But such charges are patently untrue. Environmentalism is based on science; it is, as Devra Davis, former director of the environmental board of the National Academy of Sciences, observed, "a shotgun wedding of science and law."[21] Science and technology have given environmentalists and policymakers the tools and the information to spot threats to human health and the environment caused by science and technology. For example, Ambassador Richard Elliot Benedick, a chief negotiator of the Montreal Protocol, the treaty aimed at phasing out ozone-depleting chemicals, pointed out that that treaty was made possible only by highly advanced science that projected a deadly danger well before any of its effects were visible.[22]

One of the chief functions of environmentalism, in fact, is to act as an intermediary between science and the public by monitoring scientific findings related to the environment and then calling them to the attention of and interpreting them for the media and government officials. Aside from fringe elements such as Earth First! in the United States and some fundamentalist Greens in Europe, environmentalists do not call for an abandonment of technology; indeed, they are well aware that threats to human health and welfare caused by technology often can be addressed only by a better, more benign technology. While environmentalists might be, as sociologist Lester Milbrath and others speculate, an advance guard leading the way toward a postindustrial society, they are certainly not pointing toward a post-technological society.

At the same time, however, the environmental movement has been an active, aggressive, and often successful foe of what they and their allies consider to be misapplied science and destructive technology. The political scientist Robert C. Paehlke commented: "Environmentalism can be seen as a political movement that seeks to impose on the physical sciences and engineering restraints based on the findings and judgments of the social and life sciences . . . environmentalism fundamentally shifts the purpose of science. The new science can continue to advance productive efficiency, but that efficiency must also be seen in organic and ecological rather than merely in mechanistic terms." Environmentalists seek to shift the burden of proof about the safety of technologies: they must be presumed guilty until proven innocent.[23]

In large part because of the pressures applied by the environmental movement around the world, science has been increasingly turning its attention to the effect of its own works on the natural world. Environmental journalist Gregg Easterbrook wrote that "many scientists have become converts to the environmental cause."[24] But if it was a conversion, it was caused by their own accumulating empirical findings, not by any transcendental experience. A 1990 report by the National Academy of Sciences stated: "As the pervasive effects of human activity on the earth system become clear, the world's scientists face an urgent challenge: Can they apply the scientific understanding and technology that have allowed us, for instance, to venture into space, to develop the scientific understanding necessary to address the challenges we face in protecting the global environment on our own planet?"[25] The report said that the expanding knowledge about the effects of human activity had already spurred "a new push to understand the earth . . . in one of the broadest scientific inquiries in history."[26] Virtually every scientific discipline has been pulled into the investigation of threats to the global environment and the search for solutions.[27]

. . .

For a variety of reasons, however, the intensifying scrutiny that science has been applying to the degradation of the environment has been generally slow to influence public policy. For one thing, scientific knowledge about environmental effects of human activity remains "scarce" despite the recent attention being paid within the scientific community. There is still relatively little knowledge about how natural systems such as wetlands actually work and even less known about massive global systems such as the atmosphere and the oceans.[28] One of the reasons for this scarcity of knowledge is that scientific inquiry has, over the years, tended to be concentrated in ever more narrowly focused specialties, whereas the understanding of natural systems requires a broad, cross-cutting research approach.

In the absence of firm data about the extent, causes, effects, and implications of what is happening to these natural systems, many scientists have been reluctant to make public statements on their own views, much less offer advice on policy responses. This professional reticence is reinforced by the fact that most of the scientists depend on money from industry or the government for their research—and their livelihood. Scientists who know their funding may be cut off or that they may miss out on promotion or even lose their jobs by presenting scientific conclusions that conflict with institutional policies and prejudices may not be inclined to bite the hand that feeds them. Science, like statistics and public opinion polls, can be designed to produce the results sought by those who pay for it, whether the payer is a government agency, a big corporation, or an environmental group. Even when scientists honestly believe that their conclusions reflect a strict adherence to pure science, their work can be subtly influenced by other factors such as economics, peer pressure, or personal intellectual or ideological predisposition.

Scientists are also well aware that bearers of good news, not bad, are those most often rewarded by society. As environmental journalist John Flynn noted, it is the physicists who discover new elements of matter, the biochemists who unlock the secrets of DNA, and the medical researchers who find new cures for disease who win the Nobel prizes, not the ecologists who warn that human activity is wiping out other forms of life on earth at an accelerating rate.[29] As the advanced guard of material progress, scientists have been accorded a place of high esteem in Western civilization. As witnesses to the truth that there are limits to what the earth can produce and absorb, they are unlikely to be placed on the same lofty pedestal.

The absence of complete certainty on complex issues also inhibits many scientists from giving concrete policy recommendations. David Schindler of the University of Alberta said that "scientists are more right-wing than left-wing. They're very conservative, and in fact have not over-predicted the effects of acidic rainfall, ozone depletion, and warming climate."[30] In the absence of certainty there is inevitably disagreement among scientists. Individuals are therefore reluctant to stick their necks out on policy matters for fear of bringing down the scorn of their peers—witness Jim Hansen's fate. Opponents of national and international actions to address environmental problems—the fossil fuel industry on global warming, the forest products industry on tropical deforestation and species loss, and for many years before the discovery of the ozone hole over Antarctica, the chemical industry on threats to the ozone layer—laid heavy emphasis on the absence of scientific certainty as a rationale for inaction.

As long ago as the 1860s, however, George Perkins Marsh recognized that policymakers could not wait for scientific certainty to take actions that would slow the destructive effects of human activity. "But our inability to assign definite values to these causes of the disturbance of natural arrangements," he wrote, "is not a reason for ignoring the existence of such causes in any general view of the relations between man and nature, and we are never justified in assuming a force to be insignificant because its measure is unknown."[31]

Within scientific as well as environmental circles, it is increasingly acknowledged that the magnitude of the dangers to humanity and its habitat from eroding life-support systems is far too great to wait for scientific certitude and consensus before acting to address them. The 1990 National Academy of Sciences report found that scientific evidence indicated that, because of human activities such as agricultural overuse of the land, clearing of forests, treating the atmosphere, land and water as receptacles for waste, and the way we consume energy and materials, "the global environment is undergoing profound changes. *In essence, we are conducting an uncontrolled experiment with the planet*" (emphasis added).[32]

If we are conducting such a Faustian experiment, then the dangers of inaction to control it would clearly appear to outweigh the dangers of action by orders of magnitude. If our fears prove to be exaggerated, addressing such problems as global warming and species loss would lead, at worst, to severe but temporary economic hardship, and that could be managed to mitigate suffering. If, on the other hand, the "experiment" is indeed breaking down the systems that sustain life, our generations are imposing on posterity a terrible and permanent legacy of an increasingly impoverished, inhospitable, and dangerous planet, as the Global 2000 report prepared by the administration of President Jimmy Carter suggested. The "precautionary principle," which really boils down to simple prudence, would seem to dictate a policy response to these threats now, even in the absence of conclusive scientific evidence.

But as Michael Oppenheimer and Robert Boyle contend in their book *Dead Heat*, "decisive political action" on these momentous issues cannot take place without leadership from scientists. Scientists like to remain comfortably ensconced in their empirical data, but because the world is changing so quickly, "a little more intellectual risk is required of them."[33]

. . .

There always have been, of course, scientists willing to take the risk of confronting the political status quo with their ideas and discoveries. Galileo was persecuted because his conception of the universe challenged not only the theology of the church but its political power as well. In our time, scientists such as Rachel Carson and Barry Commoner risked and attracted the opprobrium of their peers to warn of the dangers of widely used, highly profitable technologies.

In recent years the number of scientists who have been willing to step out of their classrooms and laboratories and take the intellectual risk of entering the political arena to speak out in defense of the environment has been slowly expanding around the world. Because they have been increasingly instrumental in prodding governments and international institutions to act, it would be useful at this point to acknowledge briefly a very small sampling of these socially aware, politically active scientists, although obviously only a fraction of the hundreds around the world can be mentioned here.

. . .

The threat of atmospheric and climate change has prompted a surprisingly substantial number of scientists to enter the political lists in order to persuade governments to restrain fossil fuel use, restrict the use of chemicals that are changing the composition of the atmosphere, and take other steps to limit the scope and mitigate the effects of anticipated global warming.

F. Sherwood Rowland, who with Mario J. Molina investigated the implications of the hundreds of millions of pounds of CFCs sent into the stratosphere by industry, came home from his lab one evening in 1974 and told his wife, "The work is going very well, but it looks like the end of the world."[34] The "work is going very well" part of his comment reflected a scientist's satisfaction that his research was showing concrete results; in this case that CFCs accumulating in the stratosphere would destroy the ozone molecules that protect the earth from dangerous ultraviolet radiation from the sun. The traditional scientific follow-up to such a discovery would be to publish in scientific journals, lecture at gatherings of fellow scientists, and let it go at that. But as the "end of the world" comment suggests, Rowland's concerns went beyond his own scientific achievements. In fact, he pressed for international action to limit the use of the chemicals and became a familiar figure in Congressional hearing rooms and at international forums. When he won the Tyler Prize for environmental achievement in 1983, the citation noted that his "policy initiatives led to regulations that controlled the use of CFCs as aerosol propellants in the western world."[35] His continued activism helped pave the way for the treaty to eliminate CFCs.

Sweden's Bert Bolin, mentioned earlier, also won a Tyler Prize, his citation noting that he was "a pioneer in global climate changes" and that "he helped focus international attention on the potential dangers to the world's climate posed by greenhouse gases and acid rain." Bolin was also a leader of the scientists on the Intergovernmental Panel on Climate Change, the body formed by the World Meteorological Organization and the UN Environmental Program to verify the scientific information surrounding global warming and to make policy recommendations.[36]

Stephen H. Schneider of the National Center for Atmospheric Research and Stanford University, who has studied, written, and spoken widely on climate issues for many years, is one of the more effective advocates for a policy response to global warming. An article in *Science* magazine commented that if James Hansen is a "witness" who feels he cannot keep silent in the possession of important information about climate change, Schneider is a "preacher" who invokes moral responsibility for inserting himself in the political debate. "A human being has an obligation to make the world a better place," the article quotes Schneider as saying.[37] Gordon McDonald, Roger Revelle, and Ralph Cicerone are others who played vital roles in calling the attention of the public and policymakers to the emerging and alarming body of knowledge about the greenhouse effect. Geographer Anne Whyte has served as chairman of Canada's Global Change Program and sits on the International Advisory Committee for the Environment of the International Council of Scientific Unions. José Goldemberg, a physicist who served as Brazil's environmental minister and also as secretary of state for science and technology, has done extensive research into—and advocacy for—means of reducing fossil fuels use. Goldemberg also helped slow illegal destruction of the Amazon rain forest. British meteorologist and physicist John Houghton played a central role in assessing the evidence gathered by the International Panel on Climate Change and labored to translate scientific findings into a policy response by the international community. NASA's Robert Watson played a key role in making sure that policymakers in the United States and around the world understood the implications of the ozone hole over Antarctica and the increasing deterioration of the ozone layer over the Arctic. Michael J. Oppenheimer, an atmospheric physicist and senior scientist for the Environmental Defense Fund, is probably the most influential scientist among those working on climate issues for environmental organizations.

The greenhouse issue has also animated a number of scientists to activism in opposition to policy steps to prepare for a warmer earth. In the United States, Pat Michaels of the University of Virginia is among the most vocal and persistent greenhouse skeptics, but Richard S. Lindzen, meteorologist at the Massachusetts Institute of Technology, who has a more substantial reputation among his peers, has probably been a more effective counterweight to those scientists pressing for action to deal with climate change.

In an essay on environmental science, journalist Greg Easterbrook asserted that many scientists who support "global warming alarm," such as Schneider, Oppenheimer, and John Firor of the National Center for Atmospheric Research, are "political liberals." Scientific "skeptics" such as Michaels, Lindzen, and Robert Jastrow of Dartmouth are "political conservatives," he contended. Easterbrook did not say how he determined the scientists' political orientation. In the case of Michaels, affiliations such as his connection with the right-wing Cato Institute are probably sufficient evidence of political inclination. He offered no evidence for the others except the fact that Lindzen smokes cigarettes. It is no wonder that a scientist who does not accept the accumulated evidence that smoking is dangerous would be skeptical about global warming! One wonders, however, if "conservative" is the proper term for scientists who agitate for continuing an "uncontrolled experiment with the planet."[38]

The cause of saving disappearing species and preserving the planet's biological diversity has attracted the passionate public support of many scientists, a number of them highly distinguished in their field. Edward O. Wilson, the Harvard biologist who helped develop a widely accepted method of estimating species diversity, has been instrumental in applying that method to conservation. His books have rallied many to the cause of species preservation. Thomas E. Lovejoy, a tropical biologist who serves as assistant secretary for external affairs of the Smithsonian Institution, has been for many years one of the most politically active and effective policy advocates in the United States. Among other things, he devised the "debt-for-nature swaps," a win-win technique with which developing countries are relieved of part of their external debt in return for a commitment to preserve land or wildlife. Thomas Odiahambo, a Kenyan entomologist and president of the African Academy of Sciences, has worked to reduce the use of ecologically damaging pesticides and also has spoken out forcefully on African poverty as a cause of environmental degradation on that continent. Robert Leakey, an anthropologist who headed Kenya's wildlife service, labored against heavy odds to protect that country's animal life. When political enemies forced him to resign, he was replaced by David (Jonah) Western, another distinguished and dedicated scientist.

Vandana Shiva, a physicist who founded and directs India's Research Foundation for Science, Technology and Natural Resources, works closely with the Chipko movement in northern India, where local residents, especially women, embrace trees to save them from loggers. Arturo Gomez Pompa is a Mexican ecologist who, as a special adviser to President Carlos Salinas, helped turned the attention of the Mexican government to issues of biological diversity, tropical forest preservation, and sustainable development. George Woodwell, founder and director of the Woods Hole Research Center, has long been one of the most politically active scientists in this country and often works directly with environmental groups such as the World Wildlife Fund, the Environmental Defense Fund, and the World Resources Institute. Peter Raven, director of the Missouri Botanic Garden, is a botanist who has been for many years eloquently engaged in a range of issues affecting the health of the global environment from population to biological diversity.

Dr. Philip Landrigan of the Mount Sinai Environmental Health Center and Dr. Theo Colborn, who warned that synthetic chemicals are disrupting the human endocrine system, are among the many scientists who, in the tradition of Rachel Carson, alert the public to the threats to their health from hazardous substances in the environment.

Among scientific groups perhaps none has been more active and effective in mobilizing the international scientific community for the defense of the planet than the Union of Concerned Scientists under the chairmanship of Henry Kendall, the Nobel Prize–winning physicist from M.I.T. The UCS, long a force in the struggle against nuclear testing and proliferation, made a major statement in 1992 with its "World Scientists' Warning to Humanity" (mentioned briefly in chapter 1), which calls for limits to environmentally damaging activity, improved resource management, increased economic and family planning assistance for developing countries, more aggressive steps to reduce poverty, and guarantees of sexual equality. The "warning" was signed by 1,575 scientists from sixty-nine countries.

A number of other mainstream organizations, including the International Union of Scientific Councils, the World Meteorological Organization, and the Scientific Committee on Problems of the Environment (particularly on the nuclear winter issue), have provided scientific fuel for the environmental policy debate. Medical doctors belonging to Physicians for Social Responsibility have been increasingly active in studying and alerting the public to the health consequences of pollution, toxins, and other environmental threats.

. . .

These and many other citizen-scientists who have risked taking a public position on the great environmental issues of our time have played a crucial role in narrowing the dangerous gap between science and policy. These women and men clearly have subscribed to a principle stated in Aristotle's *Nichomachean Ethics*: "Every science and every inquiry . . . is thought to aim at some good."

But the gap remains wide, particularly at the international and global levels, and the environmentalists have been only partially successful in their attempts to bridge it. Because of the continuing lack of definitive answers about complex problems, the narrow focus of much contemporary science and the lack of social science in analyzing these issues, the caution and timidity of many in the scientific community and the ideological divisions within that community; because of the powerful monied interests that have been able to block the translation of scientific discovery into remedial action; and because of a political process that is able and willing to use scientific uncertainty as an excuse for inaction, science has not fully joined the effort to meet what is increasingly recognized as a global crisis.

Because of all of these difficulties, there was a widening belief as the 1990s began that the process of translating science into policy most be broadened and democratized. Given the enormous stakes involved in addressing a deteriorating global environment, environmental science has become much too important to leave entirely to the scientists.

In their paper "A New Scientific Methodology for Global Environmental Issues," Funtowicz and Ravetz assert that "we have now reached the point where narrow scientific tradition is no longer appropriate to our needs. Unless we find a way of enriching our science to include practice, we will fail to create methods for coping with the environmental challenges, in all their complexity, variability and uncertainty."[39] The method of enrichment they propose they call "post-normal science" and others call "civic science." It proposed enlarging the participation in all aspects of scientific enterprise through "extended peer communities." In addition to the scientists with special expertise in the narrow problem area, such as atmospheric chemistry, an extended peer community would include those with scientific and technical know-how, with direct experience of and knowledge of the problems, and with an interest in the ethical and economic issues involved. Such a peer group would include not only experts but investigative journalists and citizens with a stake in a solution to the problem.

"The democratization of political life is now commonplace; its hazards are accepted as a small price to pay. Now it becomes possible to achieve a parallel democratization of knowledge, not merely in mass education but in enhanced participation in decision making for common problems."[40]

Another scholar, Kai N. Lee, director of the Center for Environmental Studies at Williams College, has proposed a strategy he calls "adaptive management" to help the world community deal with uncertainty as it gropes for policies to reverse environmental decline and to find the right path toward sustainable development. By adaptive management he means a process of continual social experimentation to find out what policies can be best employed to foster economic well-being while preserving ecological systems.

To help the global community find its way in this trial-and-error process, Lee suggests two "navigational aids." One is the "compass" of the physical and social sciences, with their tradition of research and rigorous analysis, that will point us along the right track in the direction of environmentally sustainable development. The other is the "gyroscope" of democratic debate, which, by subjecting solutions offered by scientists and adopted by policymakers to the rough-and-tumble of free political debate, ensures that we will not veer too far off course and onto the rocks of ecological or economic disaster.[41]

. . .

The modern world—or the postmodern world if that is where we are heading—cannot do without science and technology. Our civilization has become too complex, the means we have devised for our survival and comfort too intricate, and our manipulation of nature too labyrinthine to extricate ourselves from our total dependence on science and its works. And we would not want to do so if we could. The pastoral ideal of the simple shepherd living in a harmonious world of quiet nature or of the sturdy, contented yeoman farmer who depended only on his own labor and perseverance for his sustenance and happiness were largely myths. For most humans, life in prescientific, preindustrial cultures was difficult and tenuous.

But neither can we any longer permit the careless adoption of the products of science and technology without the precaution—the "gyroscope"—of democratic oversight. When science can put in jeopardy the productivity of the earth and oceans, when it can threaten to disrupt the atmosphere that shelters our planet, when it can unleash forces that can sicken and kill humans and other life, the stakes are much too high. We can no longer allow decisions about the deployment of technologies to be made exclusively by a scientific elite or by governments that listen only to that elite.

Milton Russell, then assistant administrator for policy and planning of the EPA, said a decade ago: "We are both a democracy and a technological culture. Our people will not stand to be mere victims of forces beyond their control. But if this control is unwisely applied, technology cannot flourish. Yesterday we lived in democratic societies that were not technological cultures. Today, we observe many technological cultures that are not democracies. Success in the twenty-first century will be to remain both."[42]

Chapter 7

Getting and Spending

There is no wealth but life.
—John Ruskin, *Unto This Last*

Sustainable development is not a new idea. Farmers have always known that they must save their seeds and care for their fields in order to keep their land productive and increase their yields. The French Physiocrats of the eighteenth century held that economic activity must be governed by "the rule of nature."[1] In the United States, Theodore Roosevelt's Progressive movement of nearly a century ago was based in large part on the conservation and equitable distribution of public resources—assuring, for example, that there would be a "sustainable yield" of timber from national forests that would serve future generations of Americans.

More recently, the economist E. F. Schumacher, in his highly influential book *Small Is Beautiful,* called for an "economics of permanence," in contrast to the prevailing economics of production, consumption, waste, and violence.[2] The World Conservation Strategy proposed in 1980 by the International Union for the Conservation of Nature advanced the idea of sustainable development. More than a decade ago, Lester R. Brown and the Worldwatch Society published *Building a Sustainable Society,* which proposed shifts away from economic practices that threatened the viability of the soil, water, and atmosphere that are the basis of life on earth.[3] The Founex report to the Stockholm Conference, as noted above, underscored the intrinsic relationship between environment and development. A central premise of the modern environmental movement, in fact, is that human activity—that is to say, economic activity—threatens human survival and therefore is not sustainable.

But it was not until 1987, with the publication of *Our Common Future,* the report of the World Commission on Environment and Development, that the concept of sustainable development entered—or more accurately, was thrust—into the mainstream of economic and political discourse around the world.[4]

The panel, usually called the Brundtland Commission after its chairwoman, Gro Harlem Brundtland (then former and future prime minister of Norway), was authorized by the UN General Assembly in 1983. The commission, which included members from twenty-one different countries and five continents, was charged with preparing "a global agenda for change."[5] Such change was needed because of what the commission's report would describe as the worsening "interlocking crises" of failing economic development and a deteriorating environment from the local to the global level.[6]

Current economic practices in both the rich and the poor countries were, for differing but mutually reinforcing reasons, leading down roads to disaster, the commission found. In the industrialized countries of the North, excessive and accelerating consumption of goods produced by often harmful technologies was using up raw materials and energy as well as "sinks" for absorbing pollution, such as the air and atmosphere, rivers, lakes, and oceans, at rates that could produce catastrophe in the foreseeable future. In many of the so-called developing countries of the South, the need of a rapidly growing population to provide itself with the bare necessities of life—food, water, fuel, clothing, shelter—was placing intolerable stress on the resource base that provided those very necessities. Demand for fresh water for drinking and irrigation exceeded the supply and permanently exhausted water tables. Gathering of firewood for cooking faster than trees could grow prevented woodlands from regenerating. Increasingly intensive cultivation of land to feed rapidly growing families and villages eroded the land to the point that the land could produce no food at all. Poverty itself is thus a chief cause of environmental decline, which in turn can be an insurmountable obstacle to economic development. Moreover, the poverty of the developing countries placed increasing demands on the resources of the rich—although, to be sure, the rich often declined to meet those demands.

The way out of the economic/ecological cul-de-sac, the commission determined, was to replace current approaches to economic growth with a new model, called sustainable development, which is defined as "development that meets of the present without compromising the ability of future generations to meet their own needs."[7]

The simplicity of the definition is deceptive. In economic terms it contains radical implications. The idea of meeting "needs" introduces into the neoclassical economics of industrial capitalism the alien concept of equity—equity in meeting the needs of the poor of the earth today and equity in meeting the needs of humans as yet unborn, both for economic resources and for high quality of environment.[8] Edith Brown Weiss, an expert on the legal issues of intergenerational equity, notes that "sustainability, which implies intergenerational fairness, is possible if we look at the earth not only as an investment opportunity but as a trust, passed on to us by our ancestors to be enjoyed and passed on to our descendants for their use."[9]

The commission explicitly challenged the political rhetoric that claimed that protection of the environment can take place only at the expense of economic growth. In fact, as the economists who prepared *Blueprint for a Green Economy* noted, the underlying theme of the Brundtland report "is that environmental policy matters not just because natural environments have value in themselves but because environments and economies are not distinct. Treating them as if they were is the surest recipe for unsustainable development."[10] In other words, far from being an obstacle to economic growth, care for the environment is imperative if growth is to continue. Conversely, economic growth is imperative for the preservation of the environment.

But the kind of economic growth proposed by *Our Common Future* was not business as usual. The report called for a change in the *"quality"* (emphasis added) of growth "to make it less material—and energy-intensive and more equitable in its impact . . . to maintain the stock of ecological capital, to improve the distribution of income and to reduce the degree of vulnerability to economic crises."[11] Its prescriptions implied a kind of global bargain to protect the earth. The industrialized countries would enable the developing world to lessen the destruction of local environments by a series of policy steps to alleviate poverty, including increased development aid, debt relief, access to their markets, and improved terms of trade. By, in effect, calling for a massive transfer of resources from North to South, the report also suggested that there would have to be a change in the grossly disproportionate consumption of materials and energy by citizens of the rich industrialized nations. It called for an end to the "arms culture" that was absorbing nearly $1 trillion a year that could otherwise be help fuel economic development. To uphold their end of the bargain, the developing countries would have to control their population growth; exercise better stewardship over the land, the forests, the wildlife, the other ecological resources within their boundaries; and pursue development in ways that did not degrade the global commons.

. . .

By 1987 public and political opinion had been primed for the Brundtland Report by a seemingly endless stream of bad environmental and economic news. The development process had been stopped dead in its tracks in many areas and in some, particularly sub-Saharan Africa, had gone into reverse. Stagnant economies in many of the industrial democracies contributed to the plight of the poor countries. Development failures combined with ecological decline and drought triggered a devastating wave of famine in the Horn of

Africa and elsewhere on that beleaguered continent, killing as many as a million people. In the nine hundred days the commission had been working on its report, an estimated sixty million people, most of them children, died of diarrheal diseases from unsafe drinking water and malnutrition.[12] In central Europe and in high altitudes in the United States, acid rain combined with other air pollution was killing trees at alarming rates. Toxic and nuclear wastes were piling up in many of the industrial countries, some of which were trying to export that waste to poorer nations for a price.

Meanwhile, the accumulating evidence that human activity was beginning to alter the environment on a global scale through the greenhouse effect, the destruction of the ozone layer, the rapid leveling of forests, the extirpation of species, and the destruction of topsoil had prepared fertile ground for the Brundtland message to take root. Everyone on earth was at risk, and no one country or group of countries could avert that risk alone. The unsustainable consumption patterns of the rich could and did affect the well-being of people living far away in the poor countries. And the explosive population growth and destruction of the resource base could and did affect the world the rich people lived in and, without change, would do so to a far greater degree in the future.

The World Commission report was therefore hailed North and South, East, and West as a conceptual breakthrough that could pave the way for a new approach to economic development and environmental protection.

. . .

But official support for the commission's work was in most cases entirely rhetorical. The concept of sustainable development was foreign to the economic principles of the industrialized nations, particularly to the conservative administrations of Ronald Reagan in the United States and Margaret Thatcher in the United Kingdom, which adhered with religious fervor to neoclassical economic strategies. This pure economic faith was strongly reinforced in the late 1980s and early 1990s by the collapse of the Soviet empire. The disappearance of the Communist command economies was widely hailed, particularly by conservatives in North America and Europe, as the final triumph of the current system of free-market capitalism, which, accordingly, was thenceforward beyond challenge. The thorough overhaul of the system that would be required by an economics based on the goal of sustainable development was not even to be considered.

That there were imperfections in the way market capitalism was functioning around the world, however, was obvious to anyone willing to look. In much of the South, as we have already noted, the development process was in trouble; poverty and hunger were rife. And wide cracks were starting to show even in the economic structures of the rich industrialized nations. The United States, Europe, and Japan all experienced prolonged downturns in economic activity that were widely regarded as reflecting basic structural problems rather than the usual doldrums of the business cycle. Unemployment remained at high levels in many countries. The United States, the world's biggest economy, suffered perennial budget and trade imbalances and a declining standard of living for a growing number of its citizens. The ranks of the homeless began to swell, first in the United States and then in Europe and even in Japan. The distribution of wealth, not only between the rich and the poor countries but within many of the rich countries as well, was growing increasingly inequitable. Moreover, given the concentrated economic—and political—power wielded by a comparatively few but huge transnational corporations; the industrial, agricultural and trade subsidies provided to favored economic interests by national governments; and the manipulation of global money, equities, and commodities markets, the "freedom" of the free market all too often appeared to be a polite fiction.

With the spread of crime, terrorism, and other violence; drug addiction and alcoholism; spouse and child abuse; rising suicide rates; disintegration of families; and other indicators of social alienation, it certainly would be difficult to make the case, as many observers have pointed out, that the growth- and consumption-oriented economies were enriching the quality of life for the majority of their citizens.[13]

Implicit in the concept of sustainable development and explicitly stated by a growing number of economists and social critics is the idea that the world has outgrown—in human numbers, in the size of its economy, and in the power of its technology—the economic system to which it has long been subservient. This view was stated bluntly if in somewhat hyperbolic terms by "Adam Swift," a former senior official of the British government, who wrote, "we have been ignoring a gigantic error," which was "to assume that an economic theory devised two centuries ago for a grotesquely simpler life, and a matching political theory only 50 years younger, were still vitally attuned, not only to our deeper and less creditable motives but also to the meeting of today's more complex demands." That system, he continued, "is now producing *Homo economicus,* a grotesquely distorted human being fired only by greed and false standards."[14]

To these critics the basic problem is that the economics of market capitalism—and Marxist materialism as well, for that matter—cannot deal with environmental problems because they do not even account for the existence of such problems. As the economists Herman E. Daly and Kenneth N. Townsend note, in neoclassic economics the economy contains the ecosystem, not the other way around. It perceives the economy as a closed, circular system in which labor and capital meet the perceived needs of consumers, which in turn creates more capital and demand for more labor and capital, producing a condition of continual and boundless growth. Nature is regarded as just one more subsystem of the whole economy, like agriculture and industry. Resources such as raw materials and sinks for pollution are treated as infinite.[15]

Economics is called the dismal science, but as currently practiced, it must be regarded as the myopic science (if indeed it can be considered a science at all). Neoclassical economics essentially pays no heed to the needs of future generations. With its use of discount rates and opportunity costs of capital, "the future disappears for decision purposes after a few decades."[16]

With the hindsight of modern ecological awareness, it can be seen that neoclassical economics subordinates the material world to an artifice, a series of invented principles governing much of human activity. Those rules are based on one perception of human behavior but, it is becoming disturbingly clear, not on the physical, chemical, and biological realities of the planet. Such economics are like the *Glasperlenspiel,* the elaborate, intellectually intricate "glass bead game" played by the cloistered, hermetic community in the Hermann Hesse novel. The rules were an invention that bore but a shadowy, distorted relationship to the outside world, but they dominated every aspect of the lives of those who dwelt within the fictional community. Similarly, our created economic rules may or may not assure human welfare and happiness, particularly over the long run, but they have acquired an aura of immutability and are guarded with religious vigilance by the orthodox.

The rules governing growth-dependent market economies worked well enough in a relatively empty world, with abundant resources, accessible sources of energy, and adequate reservoirs into which to dump the effluvia of industrialism and other human activity. But the exponential growth of human population and their economic activity in this century is rapidly filling up the world. A growing number of scholars are reaching the conclusion that the goal of achieving the most efficient throughput of materials and labor to achieve maximum growth is no longer appropriate and, in fact, presents a grave danger to our near future.

But as Kenneth Boulding observed, most economists still cling to an economic theory based on infinite resources, which he called a cowboy economy because the cowboy is "symbolic of illimitable plains and also associated with reckless, exploitive, romantic and violent behavior. . . . The idea that both production and consumption are bad things rather than good things is very strange to economists, who have been obsessed with the income-flow concepts to the exclusion, almost, of capital stock concepts."[17] The economy of the future, Boulding asserted, would be a "spaceman economy" because the earth is like a spaceship, with finite supplies of everything, including places to dump pollution."[18]

In a provocative essay titled "Boundless Bull," Herman Daly likened the American economy to a series of television commercials run by the Merrill Lynch brokerage firm, which shows an obviously virile bull walking across an empty beach or down an empty street or standing on an empty mesa, while a voice in the background proclaims a world of "no boundaries." The message, of course, is that investing in stocks through Merrill Lynch will put the viewer in an "individualistic, macho world without limits."[19]

But, Daly wrote,

> No bigger lie can be imagined. The world is not empty; it is full. . . . Unlike Merrill Lynch's bull, most do not trot freely along empty beaches. Most are castrated and live their short lives as steers imprisoned in crowded, stinking feed lots. Like the steers, we too live in a world of imploding fullness. The bonds of community, both moral and biophysical, are stretched, or rather compressed, to the breaking point. We have a massive foreign trade deficit, a domestic federal deficit, declining real wages and inflation. Large accumulated debts, both foreign and domestic, are being used to finance consumption, not investment.[20]

From this perspective, our current economic practices are like the sorcerer's apprentice, drowning the affluent minority of the world's people in an uncontrollable, rising flood of material goods and services while draining the resources needed to end poverty and promote development for everybody else and to assure a livable planet in the future. The dilemma for economic policy is how to break this spell without throwing our civilization into misery and chaos.

. . .

To a growing number of economists and other scholars the answer—the way off the accelerating treadmill and onto the path of sustainability—is a new kind of "transdisciplinary" economics that recognizes and incorporates the relationship between environmental systems and economic systems. A relatively new approach, although one with long antecedents, the new transdisciplinary system is called ecological economics. Robert Costanza, an active practitioner of the new economics, described how it departs from "conventional" economics in a number of crucial ways:

- Instead of regarding human tastes and preferences as the driving force of an economy fueled by limitless resources, ecological economics is based on human understanding of the constraints placed on their consumption by finite resources.
- Instead of placing economic activity in a short time frame, usually one to four years, ecological economics looks at multiple time frames, from days to eons. It places both economics and ecology in an evolutionary context.
- Instead of pursuing only economic utility to human beings, ecological

economics takes account of the interrelationship between humans and the rest of nature.

- Instead of the goals of maximizing the growth of national economies, the profits of corporations and consumption by individuals, ecological economics' goal is the sustainability of the combined ecological/economic systems.
- Instead of focusing on mathematical tools that break the economy into atomistic parts, ecological economics employs a holistic view of problems.[21]
- In contrast to conventional economics, which is inherently optimistic about the power of human ingenuity and technology to overcome all obstacles to continuous, unrestrained economic growth, ecological economics is "prudently skeptical" that such growth can be sustained. Sustainable development, according to Costanza and his colleagues, means that growth must be constrained by limiting consumption to that which can be continued indefinitely without degrading stocks of ecological capital.[22]

Most proponents of sustainable development insist that it does not require an end to economic growth, although it would compel a different kind of growth. Indeed, the Brundtland report calls for a rapid expansion of the world economy just to provide the billions of people now living in the developing countries—and their progeny over the next fifty years—a rough equivalent of the standard of living now enjoyed by residents of the wealthy industrial nations. Jim MacNeill, who served as executive director of the Brundtland Commission, estimated that to carry out its recommendations would entail a five- to tenfold increase in worldwide economic activity over the next half century.[23] MacNeill and many economists and others believe that such growth is conceivable without ecological catastrophe but only if it is growth that does not encroach on the earth's "capital," its stock of ecological resources. It would also require broad social change, including a more equitable distribution of the fruits of growth both within and among nations.[24]

But a number of economists and physical scientists are highly skeptical about the feasibility of sustainable development that entails continued physical growth of the economy, even development far short of that suggested by the Brundtland report. Herman Daly contends that "the term 'sustainable growth' when applied to the economy is a bad oxymoron." Daly stated flatly that "it is impossible for the world economy to grow its way out of poverty and environmental degradation" because the economy "is an open subsystem of the earth ecosystem, which is finite, non-growing and materially closed."[25]

To demonstrate what he calls the impossibility of growth of the magnitude envisaged by the Brundtland Commission, Daly cited an often quoted study by

Peter M. Vitousek and colleagues on the matter produced by solar energy through photosynthesis, which is called net primary production, currently preempted by human economic activity. By eating, feeding cattle, clearing forests and fields, paving over land, draining swamps, and similar activities, human beings already commandeer 40 percent of net primary production from land-based ecological systems at current population levels.[26] When global population doubles, as it is expected to do well before the middle of the next century, humans could conceivably preempt 80 percent of this primary product. Since there can be no growth that is not ultimately based on the materials and energy provided by solar energy and photosynthesis, Daly argues, it is clearly impossible for economic growth to even quadruple, much less increase tenfold. Rather than seeking sustainable growth, which, he contended, would lead us "to imminent collapse,"[27] Daly argues that future development must be based on a "steady state" or "no growth" economy, which is achieved by maintaining a constant stock of physical wealth and a stable population, with minimal throughputs of materials and levels. Whatever actual physical growth of the economy is feasible should be left to the developing countries, where it is needed to alleviate poverty and improve standards of living. In rich countries such as the United States, where economic growth is already excessive and based on stimulated wants, the drive for growth should be replaced with efforts to improve the quality of life, by providing additional leisure and a more equitable distribution of wealth.[28] But the problem of poverty, he believes, "will not be solved within environmental boundaries without population control and redistribution of wealth."[29]

Some economists, notably Nicholas Georgescu-Roegen, have asserted that, because of the iron-bound limitations set by the Second Law of Thermodynamics, even a steady-state economy is impossible and destructive and that human welfare can be served over the long run only by a declining economy.[30]

The Second Law states that heat flows from a hot to colder body and that, as a result, all physical processes tend to decrease available energy. Because of this loss the processes tend to move from a state of order to a state of disorder, called entropy. Human economic activity is based solely on the use of low-entropy materials, such as plant life, created by solar energy through photosynthesis and fossil fuels. But using that low-entropy material, usually at low levels of efficiency, converts it into high-entropy waste. Because stocks of low-entropy material on this planet are finite, and only a finite amount of the energy from the sun is accessible to residents of the earth, economic growth must inevitably slow and eventually stop, according to this perspective. Even a declining economy cannot go on forever, Georgescu-Roegen asserted.

Economic activity cannot evade the law of entropy, Georgescu-Roegen insisted. "Nothing could be further from the truth than the notion that the economic process is an isolated circular affair—as Marxists and standard economic analysis represent it."[31]

The ideas that economic activity can create wealth independent of the stock of natural capital or that growth is the goal of economic activity have long been challenged. The British economist Frederick Soddy, taking as his theme John Ruskin's view that there is no profit in exchange, told the story of a trader who sold ten pigs with a 10 percent markup and with the profit was able to buy eleven pigs. But, Soddy noted, no pigs had been created by the transaction. The extra pig was made possible by the potato skins the piglet had eaten, and the potatoes grew only because of energy from the sun.[32] "Humanistic economists" of the last century and the early part of this century, including J. C. L. Simonde de Sismondi, John A. Hobson, and Richard H. Tawney, rejected "the conventional economic wisdom that views increased production and consumption as an end itself."[33]

Today many critics argue that not only is economic growth on the Western industrialized model not feasible for developing countries, it is not feasible for the developed countries over the long run. José A. Lutzenberger, then Brazil's minister of the environment, noted that if private car ownership throughout the world equaled that of the United States or Japan, the total number of cars in the world would rise to 7 billion when total population reaches 10 billion in the early part of the next century. "This is unthinkable! The 350 million cars we have today are too much already. But if it is impossible to extend the present way of life in the overdeveloped countries to the rest of the planet, then there is something wrong with this way of life."[34]

. . .

It is probably fair to say, however, that the notion that continual economic growth is not possible is a minority view. Most economists today would agree that sustainable growth is an achievable goal. In part this reflects the fact that most economists look only at traditional measures of development and ignore nonmonetary indicators such as degradation of the resource base.

> No economist, having studied trends in GDP [gross domestic product], determined that continued growth in output cannot be sustained. No economist has concluded from a study of time series data of the prices of natural resources that economic growth is threatened by an increasing scarcity of raw materials. Indeed, many economists . . . subscribe to the view that there has been a long term downward trend in the real prices of natural resources and use this as an argument against those concerned about the planet's limited capacity to provide materials and energy for human economies.[35]

Some of these economists subscribe to a concept of "weak sustainability," which assumes that manufactured capital is a substitute for natural capital and that as long as the combined stock of the two is not diminished, sustainable growth can be achieved. ("Strong sustainability" depends entirely on maintaining stocks of natural capital.)[36]

But many ecological economists do agree that sustainable development is an achievable goal. David Pearce and his co-authors of *Blueprint for a Green Economy* argued that "the issue is often not whether we grow or not, but *how* we grow" (emphasis is original).[37] The "how" is crucial. Proponents of sustainable growth insist that there must be changes, first, in the way economic growth is measured—the way natural resources are valued, the way environmental degradation and pollution are costed (or rather, not costed)—and also in the efficiency of the use of natural capital and in the way the benefits of the expenditure of natural capital are distributed, including the way the future is discounted. Above all, economic growth must be examined to determine whether it actually promotes welfare for the mass of humanity.

Currently, the most widely cited indicator of economic growth and the most widely used in policy formulation is the gross national or gross domestic product (GNP or GDP), which simply adds up all production without considering the costs and benefits of that production or whether it contributes to national welfare.[38] For example, when land is strip-mined for coal, the value of the coal is added to the GNP, but the loss of productive land is not deducted and, of course, the aesthetic loss from the destruction of the landscape is not even considered. When people grow sick from air pollution, the cost of the medical attention they receive increases the GNP; their lost productivity, not to mention their pain and suffering, does not reduce the GNP.

Economist Salah El Serafy noted that the GNPs of both South Korea and Malaysia have grown substantially in recent years. In Malaysia, he pointed out, the growth has come from deforesting the land and selling off the timber to Japan and other nations, while in South Korea the growth has come from "hard work." Both economies have grown, but people would have to be "blind" to regard both nations as having the same economic prospects, El Serafy said and added that "depreciation of natural capital is a sin against economics as well as the environment."[39]

There have been numerous calls for a more realistic measure of national economic welfare that includes resource depletion and the negative effects of pollution. A number of countries, including the United States, are now developing such measures.

Traditional economic accounting assigns no costs to damage caused by pollution or degradation of natural resources. These have been regarded as what the British economist A. C. Pigou described as "externalities"—external to the economic process—although, as Pigou pointed out, they can and do impose real economic costs on society.[40] Because the costs of externalities are not included in the prices of good and services, they can contribute to the misallocation and depletion of resources and generally undermine human welfare. There has been an increasing demand among environmentalists, consumer advocates, and some economists that these costs be internalized. Among other things, this will entail market values being assigned to environmental amenities, such as open space and biological diversity, and to "disamenities," such as air and water pollution, and those costs would be added to the market price of goods and services.

Economists have a particularly difficult time assigning values to those things that are not part of market transactions, such as human life, crisp air, the beauty of unspoiled landscapes. Nevertheless, they try to do so—market economics requires putting a value on pretty much everything. They measure the value of human life through such means as estimating potential lifetime earning capacity and gauge the value of preserving endangered species or natural scenery or clean air through which to watch sunsets by surveying people to find out how much they are willing to pay. Critics contend that such values are not a matter of economics but of ethics. But some environmental ecologists, such as David W. Pearce, counter that such measurement can lead to better decision making and a wiser use of budgetary resources and make the economic case for protecting the environment.[41]

One approach that has been suggested for capturing environmental externalities is "green taxes" on pollution and resource depletion. The National Commission on the Environment, a private-sector initiative in the United States in the early 1990s, suggested that "a system that taxed social 'bads' such as pollution would be better for the economy and for society than the current one that penalizes social 'goods' such as wages and profits."[42] As an example it suggested a gradually increasing federal tax on gasoline to slow the increase in driving miles, a major source of air pollution, and to conserve a dwindling fossil fuel.[43] Two German scholars who have studied the feasibility of green taxes concluded that if they could be made revenue neutral—that is, having no negative impact either on the poor or on business—they could raise as much as 10 percent of the GNP, produce gains for the environment and public health, and be acceptable to industry and society at large.[44] Another suggestion is the elimination of subsidies for economic activities such as logging, grazing, and mining that deplete resources and degrade the environment.

Sustainable growth in the future, many of the ecological economists insist, will have to depend much less on the "stock" of natural capital, particularly nonrenewable resources such as fossil fuels, and more on the "flow" of such capital provided by solar energy and photosynthesis. Transportation by horses, which are fueled by grass or grain produced by photosynthesis, was more economically sustainable than transportation by automobiles, which are fueled by finite and dwindling stocks of petroleum. In the future, solar-driven automobiles could, like horses, make use of the flow of constant energy, rather than using up the world's limited stocks of nonrenewable fuel. Of course, if there were as many horses as there are cars today, our city streets would be chin deep in manure. (But then there would be less need for chemical fertilizers.)

. . .

By the late 1980s the effects of the Brundtland report and the continuing pressure applied by environmental groups around the world, reinforced by accumulating evidence that all was not well with the global economy and environment, had begun to nudge some of the centers of economic power, such as the multilateral development banks, the transnational corporations, and even trade ministries and the General Agreement on Tariffs and Trade (GATT) onto a new course if only by a fraction of a degree.

The Bretton Woods institutions, the International Bank for Reconstruction and Development (World Bank), the International Monetary Fund, the International Development Agency, and other multilateral financial institutions were created by the Allied powers in 1944 to help reconstruct a devastated Europe and Japan and restore a functioning global economy in the postwar years. Within a relatively few years the focus of these institutions shifted to development assistance to the nascent economies of the Third World.

By the 1980s, however, these institutions, particularly the World Bank, were under sharp attack by environmental groups around the world, and many governments, including lender as well as borrower nations, joined in the complaints.[45] They pointed out that the loans tended to support huge infrastructure projects such as massive dams and big airports or to stimulate rapid development of the industrial, agricultural, and forestry sectors of the economies of the borrowing nations. Critics such as Roger D. Stone and Eve Hamilton charged that because little or no heed was paid to the environmental consequences of those projects they frequently caused massive environmental damage:

> Large new dams that created electric power and opened new lands for cultivation also destroyed forests, spread waterborne diseases, caused siltation, and forced large numbers of indigenous people to abandon their homes. Development planners measured tropical forests in terms of broad feet rather than biological diversity content, and saw the use of farm chemicals mainly as a way to increase yields. Tropical deforestation to make way for cattle pastures, generally the least efficient usage of the rain forests, to this day has its devoted advocates in development circles. The high discount rates the development agencies customarily use for cost-benefit analysis mean that environmental losses carry almost no weight.[46]

The environmental destruction caused by such projects frequently eroded the productivity of the resource base, thus restraining rather than promoting the development process.

The repayment of debt to the multilateral institutions, as well as to private lenders in the North, also proved to be an increasingly heavy burden to the developing nations. The demands of debt service grew so great, in fact, that many nations stripped the land of timber, cultivated erodible soils, and extracted minerals at a pace and in a manner that did harm to the environment in order to raise the needed foreign exchange.

The World Bank, as the preeminent lending institution, came under particularly harsh criticism. Bruce Rich, a lawyer for the Environmental Defense Fund and unrelenting gadfly of the bank, accused it of "systematic environmental negligence." The bank had been making loans for environmental projects such as reforestation, but as Dr. James Lee, the bank's environmental adviser conceded in 1986, those were "not typical projects of the bank. Traditionally our projects are more socio-economic aimed at earning more foreign exchange and creating more employment."[47]

In response to the criticism the bank began to address environmental issues with a somewhat higher degree of interest and investment. Barber Conable, president of the bank in the late 1980s, said that it must "balance growth with environmental protection."[48] Instead of financing a few scattered environmental projects, it began to integrate environmental concerns into large lending projects. It also created a new, high-ranking post to help guide the bank's environmental policies and recruited Mohamed El-Ashry from the World Resources Institute, an environmental research and policy group, to fill the post. Shortly after joining the bank, El-Ashry noted that, while the bank had regarded protection of the environment a desirable goal, it was seen as separate from and secondary to the development process. In the future, El-Ashry said, it would be the task of the bank to "integrate the environment with development" and to consider the "quality of development."[49] El-Ashry later was named head of the Global Environment Facility, a new fund created by the UN to deal specifically with environmental problems that threatened the entire planet: global warming, the destruction of the ozone layer, loss of biological diversity, and the degradation of international water resources, as well as issues such as deforestation and desertification that are caused by land degradation. The program is jointly operated by the World Bank, the UN Development Program and the UN Environment Program.

The World Bank and other multilateral development institutions were trying harder. In 1990, for example, the bank began to issue an annual progress report on its environmental activities. In its second report it stated, in somewhat self-congratulatory terms, that "the speed with which environmental concerns have been systematically integrated into the mainstream of the Bank's work . . . is due in part to the mounting evidence that continued environmental degradation threatens the attainment of the Bank's main objectives: reducing poverty and promoting sustainable development."[50] Nevertheless, well into the 1990s, funding of massive projects that destroyed ecosystems and displaced indigenous people continued to draw heavy criticism.

. . .

In the late 1980s there was some soul-searching among the rich industrial nations about levels of official development assistance but little effort to loosen purse strings. As noted earlier, only a couple of northern European countries had reached the goal of providing 0.7 percent of GNP for development aid, and most of the rich nations, including and especially the United States, were well below that level. In 1989 total U.S. bilateral and multilateral aid amounted to 0.15 percent. By contrast, the United States invested as much as 3 percent of GNP annually to support the Marshall Plan for western Europe in the years after World War II. Collectively, the nations of the Organization for Economic Cooperation and Development (OECD) set aside 0.33 percent of GNP for development assistance in 1989, less than half the UN target.[51]

A major reason for this shortfall in development assistance was that most of the industrialized countries were in or entering a period of economic sluggishness and rising unemployment. In at least some of those countries, however, there was also an almost palpable weariness and disillusion with the development process, particularly with the waste stemming from corruption and internal conflicts in many of the recipient nations. But these failures to reach the goals of unilateral and multilateral development aid were one of the key sources of the environment/development crisis.

. . .

The Brundtland report also prompted a fresh look at the way international trade affected the global environment. Environmental consideration loomed large in the debate in the United States over the North American Free Trade Agreement (NAFTA), and for a while there was some talk that the Uruguay Round of trade negotiations would lead to the "greening of GATT."

But where it intersected with environmental policies, trade proved to be an even more intractable dilemma than the environment-development puzzle is for the multilateral lending institutions. By 1990 world trade had reached an annual rate of $3.5 trillion, almost one-fifth of total global output and eleven times the trade volume in 1950.[52] So immense a level of economic activity, affecting virtually every corner of the globe, could hardly fail to affect the environment in all of its aspects.

One view of this effect, expressed in the 1990 GATT report "Trade and Environment," is that expanding trade can help solve environmental problems by promoting economic growth and thus reducing the poverty that often is the underlying cause of degrading natural systems in the developing countries and by helping finance antipollution measures. But as the World Wide Fund for Nature pointed out in a critique of the report, GATT does not distinguish between growth that is good for the environment and growth that is bad for the environment.[53] As Hillary French of the Worldwatch Institute noted, the wrong kind of trade can intensify environmental problems: "It plays a central role in deforesting Malaysia and accelerating the extinction of plants and animals in Cost Rica. It exacerbates climate change by increasing the energy requirements of goods transported over long distances." Moreover, as already noted, poor terms of trade can deepen poverty by keeping the prices poor nations receive for their exports relatively lower than the prices they must pay for their imports from the rich countries.[54]

A critical concern of environmentalists about trade policy is that it can be used as a weapon to force the lowering of environmental standards in the name of promoting freer trade. As a case in point, they frequently point to the decision by a GATT panel that upheld a complaint by Mexico and declared that a U.S. law that bars the sale of tuna that are caught along with dolphins—which are frequently found with yellowfin tuna and are accidentally caught and killed in large numbers by the tuna nets—as an unfair restraint of trade. The environmentalists have been particularly alarmed about GATT's effort to harmonize environmental standards around the world, fearing that such a process would force countries to lower their standards to "the lowest common denominator."[55] It was in large part this fear that led some U.S. environmental organizations to join with labor unions, which were concerned about the migration of jobs to low-wage Mexico, in opposing Congressional ratification of NAFTA in 1993.

But U.S. environmentalists were split over NAFTA. Some of the national organizations supported the treaty in the belief that it would produce a rising standard of living in Mexico that could support more stringent environmental regulation.

Free-trade enthusiasts believe that the expansion of world trade is so significant that other goals, such as cleaner environment, should not hinder its progress. But as legal scholar Edith Brown Weiss commented, "Trade is not an end in itself; rather, it is a means to an end. The end is environmentally sustainable economic development."[56]

Weiss also pointed to a critical difference in the "two cultures" of the environmental and the trade community. Environmental decision making, she noted is generally an open, democratic one, and environmental organizations seek wide citizen participation in the drafting and enforcement of environmental regulations. In contrast, the relationship between the trade community and governments is more closed and shuns the "public eye."[57]

The Uruguay Round of trade negotiations, which included, among other things, an agreement to replace the GATT with a World Trade Organization, ended without any meaningful greening of world trade policies. This failure pointed to continuing conflict between free traders and environmentalists, particularly in the United States. The U.S. Greens were not able, however, to block Congressional approval of the revised GATT treaty in 1994.

. . .

Private business and industry are the primary global engines of production, job creation, trade, economic growth, and, potentially, of economic development for the poorer nations of the world—all the more so with the disappearance of the centralized command economies. They are also the biggest consumers of natural resources, sources of pollution, spreaders of toxic substances, and as we have seen from incidents such as the Bhophal tragedy, sources of environmental catastrophe. And with the emergence of the great transnational corporations—companies with operations around the world—over the past half century, business and industry represent concentrations of economic power rivaled only by a handful of the largest industrial nations.

As mentioned in an earlier chapter, the fifty biggest transnational corporations account for about half of the annual direct private investment around the world each year and together employ over thirty million workers. Many of these companies, such as Exxon, Mitsubishi, Nestlé, Citicorp—have annual revenues and assets that exceed the GNP of most of the countries in the world. They control not only production but advertising and information channels as well and are able to sway consumer buying patterns; tobacco companies, for example, persuade poor people in developing countries to take up smoking.[58] They exert heavy influence over global money and commodity markets—and over most aspects of government policy in virtually every nation.

For many years after the onset of the environmental awakening, corporations dug in their heels against efforts to fight pollution and to husband resources and in general, against any limitations on their ability to operate where and how they saw fit. As the futurist Hazel Henderson commented, most corporations continued to adhere to traditional "flat-earth" economics based on "ever more complex, capital-intensive, socially and economically disruptive technologies."[59]

Gradually, however, the realities of environmental degradation and the response by environmentalists, governments, and consumers began to penetrate even into well-insulated corporate executive suites and board rooms. The change began first in the United States, where public demand prompted the enactment of serious of far-reaching environmental laws that placed new demands and restrictions on corporate activities. At first, business and industry fought to roll back environmental restraints but gradually realized that environmentalism was not going to vanish and they would have to obey the law. Soon a growing number of companies were discovering that curbing pollution and saving resources were good business practices, saving money and improving their competitive posture. This emerging view was expressed in 1990 by Nicholas L. Reding, corporate executive vice president of Monsanto Company. In a speech about a "shifting industrial paradigm," Reding said that changing corporate practices to meet environmental needs can cost "big dollars" and weaken a company competitively in the short term. "But if you're in business for the long term, corporate environmentalism makes excellent business sense. Ultimately, it means you will have more efficient processes, vastly reduced disposal costs, and broader community support."[60]

This "shifting paradigm," however, took place much more slowly in Europe and Japan; and in the developing countries, concentrating on mere survival, environmentalism was not even an issue. Some transnationals shifted operations to these countries to avoid environmental regulation, and some made Third World countries a dumping ground for wastes too unacceptable and expensive to dump at home. These giant corporations also were frequently charged with draining the profits from the exploitation of natural resources in developing countries.

Following the publication of the Brundtland report, pressures on the transnationals to be better global environmental citizens intensified markedly. The UN Center on Transnational Corporations developed a "Criteria for Sustainable Development" as a guide to corporate behavior.[61] Environmental groups and development nongovernmental organizations in the South as well as in the North stepped up their scrutiny and criticism of corporate behavior. And these powerful institutions, or at least some of them, began slowly to respond. In an earlier chapter we noted that the International Chamber of Commerce adopted a "Business Charter for Sustainable Development" in 1990. That same year, Stephan Schmidheiny, a billionaire Swiss industrialist acting at the behest of Maurice Strong, formed the Business Council on Sustainable Development. Strong, who had been named secretary-general of the UN Conference on Environment and Development, wanted to provide the UN with a business perspective on the issues. The council, composed of chief executive officers of fifty big transnational corporations, produced a book-length report called *Changing Course* two years later.

Considering its source, *Changing Course* was a radical document. It was essentially a road map for business and industry to follow in creating conditions for sustainable economic growth and called for sweeping changes in the way corporations capitalize their operations; acquire raw materials; produce, price, and market their products; and market their wastes. It also called for firm national and international regulations to deal with serious threats to public health and safety and to reinforce market initiatives to protect the environment. It proposed action by corporate leaders to address such "alarming trends" as excessive population growth, degradation of natural resources, and the gross imbalances caused by the concentration of economic growth in the industrialized countries. Most surprisingly, the report urged corporations to consider the needs of future generations, a difficult task for corporate leaders whose jobs depend on annual profits. It advised companies to abandon the wasteful "arms culture" from which many businesses profit and to adopt "full cost pricing" to reflect the costs their operations and products impose on the environment.[62] As Schmidheiny noted in a 1991 speech, "The multinational companies are among the very few vehicles available with the global scale, size, and organizational capacity to confront international environmental and development problems. The real challenge to business is how to use these capabilities and resources in a responsible and sustainable way."[63]

It is clear that by the beginning of the last decade of the twentieth century the message of sustainable development was getting through to the giant corporations. But as Irving Shapiro, former chief executive officer of E. I. Du Pont de Nemours & Co., cautioned: "Where the environment is on the corporate agenda depends on the public. If the public loses interest, corporate involvement will diminish."[64] And in fact, following the ascendance of the Republican right in the 1994 Congressional elections, many American companies joined in the effort to roll back the progress made in protecting the environment.

. . .

Broad public awareness and participation around the globe is quite clearly the sine qua non for creating a workable system for sustainable development. As the Spanish economist Juan Martinez-Alier has pointed out, "The economy and the ecology of humans are embedded in politics." But, he added, those who make political decisions today usually fail to speak for "three to four billion poorest members" of the human race who are alive today and assuredly do not represent the interest of future generations.[65]

Traditional economists argue that the marketplace itself affords the broadest possible forum for representation on economic issues because everybody has an opportunity to express herself or himself through personal market decisions. Robert W. Fri, president of Resources for the Future, a Washington-based research and policy group, stated that "the neoclassical criterion rests on the idea that everyone gets a vote; that is, individuals can choose to allocate their time and resources to maximize their own well-being." Fri conceded that not everyone gets an equal vote because the poor have too few "dollar votes." But, he added, that does not mean mainstream economics should be replaced; all that is needed is "an enlarged framework for economic analysis."[66]

But Mark Sagoff, a scholar who looks at the intersection of philosophy and public policy, noted in responding to Mr. Fri that people often choose to allocate their time and resources for reasons other than to maximize their own well-being, including ethical and ecological reasons. Therefore, "institutions of democratic government, rather than economic models, are best able to incorporate ethical, aesthetic, cultural, and other principled concerns into the formulation of environmental policy."[67]

By the late 1980s the voices of democracy were clearly making themselves heard in the environment/development debate across the world. Environmentalists and their fellow activists in the independent sector demanded, with increasing force and clarity, that the earth's needs be built into economic policy. They insisted that the pleas of peoples in the Third World for economic as well as ecological justice be heeded.

Increasingly, those voices were penetrating the sealed chambers where international diplomacy is conducted.

Chapter 8

The Greening of Geopolitics

We have it in our power to begin the world all over again.
—Thomas Paine, *Common Sense*

Gathered in an nondescript office building in downtown Montreal in the early autumn of 1987, delegates of governments from around the world took a step without parallel in the annals of diplomacy. Poor and rich nations, North and South, East and West agreed first to freeze and then reduce the production and use of chlorofluorocarbons (CFCs), chemicals of great industrial utility and prodigious commercial profitability. They did so because scientists had told them that the chemicals were probably destroying the stratospheric ozone that shielded the earth's surface from the sun's destructive ultraviolet radiation and that such destruction would sharply raise risks to human health, to crops, and to wildlife.

With the treaty, the participating nations tacitly acknowledged that they faced a new kind of danger to their security, a threat posed not by a hostile power or by domestic insurrection but by a commonplace industrial chemical. The governments that agreed to the pact, known as the Montreal Protocol, in effect yielded part of their jealousy guarded sovereign power by placing limits on economic activity within their borders for the sake of international cooperation to protect the global environment. Moreover, the international community acted without any tangible evidence that the chemicals were actually harming humans or nature; they did so solely on the basis of still evolving models—scientific projections that such damage would take place in coming years if no mitigating action were taken.[1]

Two years later, reacting to increasingly definitive evidence that the destruction of the ozone layer represented a grave peril to humanity, the parties to the treaty, meeting in London, agreed to a total ban on the production and use of most—but not all—of the valuable but dangerous chemicals. In an equally significant step, the industrialized nations, after some initial resistance by the Bush administration in the United States, pledged financial support for the developing countries, to help them acquire the technology needed for switching to alternatives for the offending compounds, which were ubiquitously used in refrigeration and insulation, as cleaning solvents for electronic parts, and in many other applications, many of them vital in the development effort. For the first time the governments of the rich countries conceded that they had a vital interest in assuring that the poor countries would participate in safeguarding the physical world they both shared. For once they were willing to provide the additional development assistance that the nations of the South had been insisting they must have to address environmental problems.

Mostafa K. Tolba, the director of the UN Environmental Program, who had played a leading role in getting the international community to focus on the ozone threat, called the Montreal agreement "the beginning of a new era of environmental statesmanship."[2] Ambassador Richard Elliot Benedick, a chief U.S. negotiator in Montreal, said that the treaty "may be the forerunner of an evolving global diplomacy, through which nations accept common responsibility for stewardship of the planet."[3]

. . .

Before Montreal there had been, as we have seen, a substantial number of multinational treaties already negotiated that pledged individual nations to cooperate for the protection of the global commons, including several that required governments to at least restrain claims of sovereign rights. Between the 1972 Stockholm Conference and 1990, sixty-seven multilateral environmental treaties were concluded.[4] Of these, perhaps the most notable was the Law of the Sea Treaty, signed by many nations (not including the United States) in 1982. Agreement had been reached after years of difficult negotiations conducted, over their latter stages, by Ambassador Tommy Koh of Singapore, a brilliant young diplomat who would later preside over the Preparatory Committee for the Earth Summit in Rio. Described as "the largest and most complex international negotiation ever held,"[5] the treaty bound signatories to observing limits on their exploitation of a shared global resource—the oceans—while also spelling out the rights of individual nations to use of that resource. The nations that signed the treaty—a majority of the international community—agreed to limit their freedom to use the common resources of the oceans as well as to accept a legally defined limitation of their territorial waters.

Significant as it was, however, the Law of the Sea Treaty did not impinge deeply on traditions of national sovereignty. As defined by Grotius, the sixteenth-century legal scholar, the sovereign powers of a nation-state are confined to the territories it controls, which generally means within its own borders, except in time of war. But the oceans are the *res communis*, the common property of all people, with the benefits to be shared by all. Thus, most of the maritime rights governments surrendered were rights that, by international legal custom, had belonged to all nations in common.[6] In fact, the treaty created a new legal concept: the extension of exclusive economic zones to nonsovereign territory.

The ozone treaty, in contrast, was an implicit recognition that no nation, no matter how powerful or isolated, could defend itself from global environmental threats by exercising its sovereign powers, even within its own borders. A puff of CFCs from an aerosol foam can of shaving cream squirted in Tokyo can contribute to dangers of skin cancer in Chile, crop loss on the Great Plains of North America, and destruction of krill in Antarctica. As Caldwell has pointed out, those threats could not be averted by competition as usual among nations but only by cooperation among nations—in effect a sharing of sovereign powers.[7]

Governments, pushed by environmentalists, had been slowly coming to recognize the interdependence being imposed on nations by the need to preserve the global environment, not to mention the global economy. Pollution drifting across national borders in central Europe and North America, shipments of hazardous materials by sea and air, the deterioration of fisheries, the rapid disappearance of forests and plant and animal species were among the issues that international diplomacy found itself addressing in the years after Stockholm.

Until the late 1980s, however, the environment was on the far periphery of geopolitics, forced to defer at every turn to foreign policy regimes focused on traditional economic and military security considerations. With the agreement in Montreal and the public attention focused on global warming during the long, hot summer of 1988, the environment made its first appearance in the center ring of international politics. Along with the dangling sword of nuclear holocaust, the growing evidence that human activity was altering the climate and atmosphere in ways that could threaten the well-being, if not the life and health, of every human being on the planet was giving a sobering new meaning to the idea of "one world."

As the last decade of the twentieth-century approached, national leaders and international organizations were scrambling to hop onto the environment and development bandwagon. In December 1988, as the Cold War was winding down, Soviet leader Mikhail Gorbachev made what seemed at the time a startling address to the UN General Assembly in which he called for "a new world order" that would entail, among other things, "restructuring the world economy and protecting the environment." In his speech he alluded to environmental threats more than twenty times and repeatedly warned that poverty in the Third World was a threat "to all mankind." He called for cooperation among all governments and an expanded role for the UN to end wars and regional conflicts and to halt "aggression against nature" and the "terror of hunger and poverty."[8] Three months earlier, Soviet Foreign Minister Eduard A. Shevardnadze had told the General Assembly that the threat to the global environment is "equal to that of the nuclear and space threat" and called for the creation of an "international regime of environmental security."[9]

On that same day, Margaret Thatcher, the conservative prime minister of the United Kingdom, who had long taken a skeptical view of environmental problems and a dim view of environmentalists, said in a speech to the Royal Academy in London that protecting the balance of nature is "one of the great challenges of the late 20th century."[10] Then Vice President George Bush, who had faithfully tried to carry out President Ronald Reagan's efforts to dismantle environmental regulation in the United States, proclaimed during his own successful campaign for the presidency that he would be the "environmental president" and pledged to convene an international conference on the environment. And in July 1989 leaders of the seven major industrialized democracies gathered in Paris for what French president François Mitterrand dubbed the Green Summit. For the first time the global environment and its links to poverty and development were given a prominent place at a meeting of world's most powerful leaders. The communiqué issued at the end of the summit described an "urgent need to safeguard the environment for future generations" and asserted that "protecting the environment calls for a determined and concerted international response and for the early adoption, worldwide, of policies based on sustainable development."[11]

. . .

Clearly, something other than diplomacy as usual was going on. Gro Harlem Brundtland, again prime minister of Norway, told an interviewer that "a revolution is taking place in 1989. For those of us who have been pressing ecological concerns for 15 years, it is astonishing to see the leading political figures of the world compete over who holds the most important meetings on the ozone layer or the greenhouse effect. It's fantastic."[12]

In part, this sudden "greening of geopolitics" as U.S. environmental activist Rafe Pomerance dubbed it, was made possible by the collapse of the Soviet Union and the fading armed confrontation between East and West. Prime Minister Brundtland noted that "already, a new awareness of global ecological interdependence is filling the political space which used to be occupied by divisive Cold War concerns."[13]

The end of the East-West conflict removed, at least for the time being, the threat of a nuclear confrontation. But as Gwyn Prins, director of the Global Security Program at England's Cambridge University pointed out, nation-states were seemingly incapable of protecting their citizens from the dangers of polluted and poisoned air and water, from changes in the atmosphere and climate, and in the poorer countries, from resource scarcity.[14] The United States, for all of its military and economic power, could not by itself avert destruction of the ozone shield or prevent disruptive changes in climate. The remoteness of small island nations like the Maldives could not save them from being inundated by rising seas caused by the release of industrial gases in countries many thousands of miles from their low-lying shores. The inability of degraded environments to provide sufficient food and water exacerbated political and cultural tensions in places as diverse as Haiti and Rwanda, all too frequently leading to civil war and the slaughter of innocent civilians. The industrialized nations of western Europe and North America were increasingly flooded by immigrants fleeing from lands where too many people were competing for too few resources, sometimes disturbing the domestic tranquility of the host countries.

In his nine-hundred-page book *Diplomacy,* former U.S. Secretary of State Henry Kissinger mentions the environment just once. Early in the volume he includes environmental matters in a list of issues, including communications, population, the world economy, and nuclear proliferation, that are tending to globalize international relations.[15] But that is far from the point of his book. Quite the contrary. Diplomacy always has reflected the self-interest of nations and has relied more on competition than cooperation, and there is no reason to believe that will change, he asserts. In fact, the disappearance of ideological conflict or a strategic military threat after the Cold War has created a world resembling the European state system of the eighteenth and nineteenth centuries, which was dominated by the *realpolitik* of sovereign states pursuing their own interests through the exercise of national power. In the struggle for the "new world order," power has become more diffuse. But America's survival, he argues, will depend on its ability to serve its own *raison d'etat*—the interests of the state—by whatever means necessary. Instead of reaching toward a new world order, Kissinger appears to be saying, we are returning to the age of Metternich.[16]

It is probably a fair guess that many of today's practitioners of international relations share Kissinger's view of diplomacy as the exercise of competitive power politics among nations. But the rise of green diplomacy in the latter part of the 1980s seemed to reflect something different: a growing awareness of a new *realpolitik* that must be addressed not by competition but by cooperation and not by unilateral exercise of sovereign power but by pooling that power to confront the complex array of environmental and economic problems that threaten all nations. What, after all, could be more real than poverty and hunger, disease caused by choking air and tainted water, people fleeing across national borders seeking sanctuary from survival-of-the-fittest brutality provoked by scarcity, or a rapidly warming planet and dangerous levels of radiation from the sun?

As Jessica Tuchman Mathews, then a vice president of the World Resources Institute, wrote in the spring 1989 issue of *Foreign Affairs,* traditional notions of national security are a "poor fit" with the efforts to address the emerging global environmental problems. "Environmental strains that transcend national borders are already beginning to break down the sacred boundaries of national sovereignty."[17] Norman Myers put the case more colorfully: "All nations are in the same boat, and it is becoming an environmental Titanic."[18]

The politics of traditional "realism," which divides the world into distinct territorial units, each with its own interests and each acting on its own, tends to ignore the potential of environmental problems to create violent conflict among nations, contends Canadian scholar Thomas F. Homer-Dixon. Internally, he adds, the friction caused by environmental stress can erode democratic institutions and fracture societies.[19] He traces civil unrest and continued insurrectionary movements within the Philippines, for example, to the economic distress and social upheaval caused by rapid deforestation, soil erosion, and population growth that created "powerful discontent" in rural areas of that island nation. As an example of potential external conflict between nations he points to the hostility between Syria and Turkey over the diminishing water resources provided by the Euphrates.[20]

Environmental causes of armed conflict have, of course, been a theme of history as long as there has been a history. More than six thousand years ago, the ancient records tell, city-states in what we now call the Middle East were at war over already scarce water resources.[21] Today the same scarcity continues to be a potential flashpoint in that region. Beyond scarcity, however, the growing awareness that the interlocked effects of economic and ecological decline can have dangerous destabilizing effects among and within nations is forcing governments and scholars to reconsider conventional approaches to national security, including military security.

No one is suggesting that military vigilance be abandoned. The world is still a dangerous place, and traditional national interests still must be defended by traditional means. But by the late 1980s it was apparent to experts on security issues, even those steeped into the arms culture of the post–World War II era, that an expanded notion of security is required by the realities of a global environmental and economic crisis.

Senator Sam Nunn, the hawkish chairman of the Senate Armed Services Committee, in a 1990 speech on the Senate floor, listed post–Cold War threats to U.S. security, including chemical weapons and missiles in the Third World, radical Moslem fundamentalism, global terrorism, and narcotics trafficking. Then he had this to say:

> I am persuaded that there is also a new and different threat to our national security emerging—the destruction of our environment. The defense establishment has a clear stake in countering this growing threat. I believe that one of our key national security objectives must be to reverse the accelerating pace of environmental destruction around

the world. . . . America must lead the way in marshalling a global response to the problem of environmental degradation, and the defense establishment should play an important role.[22]

Nunn then went on to suggest that the defense and intelligence communities could muster a number of resources to help meet the environmental challenges facing the nation and the world, including their monitoring and analysis capabilities to track ecological changes in the air, land, and water; to spot emerging trends and problems; and to model consequences and responses.

Admiral Sir Julian Oswald, the First Sea Lord of the United Kingdom, saw a need for an even more sweeping redefinition of military security. In a world joined by an increasingly shared economy and pervasive information technology, the military can no longer be the predominant source of power that provided the foundation for five hundred years of European expansion, he asserted. This reality dictates that the concept of security be expanded "to place national and personal self-interest in a global context as well as within that of a sovereign state." In that context, all nations have a stake in "environmental security," and military forces may increasingly be deployed to defend that global security. Not only would the military engage in a precautionary role of monitoring and research but could also be called on to carry out its traditional "coercive" function to protect the global commons from destruction and to enforce international environmental treaties.[23] Sir Crispin Tickell, former U.K. ambassador to the United Nations, even foresees a world police force operating under the authority of the Security Council "to compel environmental rectitude," although he concedes the thought is somewhat distasteful.[24] And Senator Al Gore said in 1989 that the imperatives of collective security will impose in the not too distant future "a new 'sacred agenda' in international affairs: policies that enable the rescue of the global environment."[25]

Within the U.S. Defense Department itself, environmental considerations began to permeate strategic thinking. A special report by the Strategic Studies Institute of the U.S. Army War College observed that "the change in the international arena since the end of the cold war has given rise to an entirely new approach to viewing U.S. security interests, and a recognition of environmental factors in international stability and the onset of conflict" and recommended that the Defense Department "proactively address environmental issues."[26]

The new diplomacy of sustainable development, finally, implied an acknowledgment that no nation could increase its security at the expense of another nation; security could be strengthened for one nation only if it was strengthened for all.[27] As such, it signaled a sharp break with the survival-of-the-powerful *realpolitik* of international relations.

. . .

The new politics of environment and development is already making significant alterations in the dynamics of North-South relations. Those dynamics are increasingly central to diplomacy, as tensions between the rich and poor countries and the collapse of social cohesion within many of the poor countries have replaced the East-West conflict as the chief threat to global peace and stability.

Both the rich industrialized countries and the poor developing countries were coming to see their own interests tied to halting and reversing the loss of resources and the degradation of ecological systems, albeit often for differing reasons. As pointed out by the Centre for Science and Environment in New Delhi, "While the rich and well fed are more interested in the environment because they want to secure their future, the poor and dispossessed, caught in a daily struggle to survive, are more interested in the environment because they want to secure their present."[28] Development is not seen as a vital issue in the North except as it promotes the building of new markets and access to raw materials. In the South environmental problems pale before the immediate suffering imposed by poverty and its attendant evils.

Nevertheless, mutual dependence on and responsibility for the global environment has underscored for governments of both North and South that, despite the deep gulf that divides them, they have a convergence of interests. Coal burned in Europe or North America is leading to global warming that will affect the citizens of Africa and Latin America. The deforestation of Brazil and Malaysia and Indonesia not only would exacerbate the warming problem but destroy the biological diversity that should serve current and future generations of people in every part of the world.

This common interest provides a powerful new incentive for the industrialized countries to help speed the process by which the poorer countries can reach a level of development that would enable them to stabilize populations and protect their resource base. It also encourages the rich countries to transfer benign technologies to the South to help reduce the environmental impact of development on the global commons. In the developing world, meanwhile, recognition that environmental protection is in their self-interest makes their governments less likely to regard diplomatic efforts on behalf of the environment as disguised neocolonialism or an obstacle placed deliberately on their path toward progress.

In the past the development assistance process had been one of inherent inequality—the poor countries dependent on the ostensible generosity of the rich. No matter that aid programs chiefly served the economic and security interests of the industrialized nations. The very concept of aid implied a paternalistic relationship. Donor countries sought to dictate not only the terms and conditions of assistance but, in effect, the kinds of societies development would create. Such inequality inevitably created resentment and increasingly soured the North-South relationship. Efforts by donor countries or multilateral institutions to superimpose environmental requirements on loans or grants was assailed by many in the Third World as "green imperialism."[29] Lawrence E. Susskind of M.I.T. commented:

> The North-South split is often portrayed as a battle over money and technology, but there is more to this conflict than economic and scientific ascendancy. . . . The injustice of cultural hegemony (that is, the overwhelming impact of Western culture and forces of modernization on economically dependent countries) undergirds the development assistance and technology transfer debates. . . . These debates mask the real source of conflict, which is a fundamental difference in how the nations of the North and the South think about progress.[30]

Southern governments are wary about international negotiations on environmental issues because they frequently do not have the technical expertise to participate fully in the debate and sometimes do not even have the funding to send a full delegation to the diplomatic conferences.

Now, however, the South has something that the industrialized countries are discovering they want and need: cooperation in safeguarding the earth's life-support systems. It gives the developing countries a new, potentially valuable bargaining chip to throw on the table in their diplomatic confrontations with the North over environment and development. As Emil Salim, Indonesia's state minister for environment and population, pointed out at an interparliamentary conference in 1990, "the extension of environmental cooperation and the provision of development assistance would no longer be an act of altruistic generosity or of charity. Instead it would be an aspect of a pattern of international relations based upon a new perception of national self-interest and upon the realization that we shall either survive together or perish separately; that unless the weak get stronger, all will become weaker and even the strongest won't survive."[31]

The developing countries felt, therefore, that they were in a position to exact economic concessions for taking steps to protect the environment that might impede their economic progress. Why, they asked, should they accept the prospect of continuing poverty in order to promote the common welfare of the entire planet? Why should they not demand the equity in economic relations with the North for which they had long been unsuccessful supplicants in return for cooperation in preserving the global environment? How, they asked, could they be expected to meet international environmental standards without the financial, technological, and trained human resources necessary to do so?

For years the developing nations had unsuccessfully been seeking to narrow the huge gap in wealth between them and the industrialized nations through the establishment of a "new international economic order" that would increase and stabilize prices for their commodities, reduce their debt burden, give them access to advanced technology at prices they could afford, and give them new and additional funding from the North for projects to protect the environment.[32] Now there were powerful incentives for the North to respond to those demands.

A more contentious issue was raised by the diplomats and activists of the South when they questioned why their countries should be asked to make sacrifices for the sake of the global environment when the voracious consumption patterns of the North are responsible for most of the serious ecological threats to the planet, when the 25 percent of the global population living in the industrialized nations consume 80 percent of the world's resources.[33] Why should they refrain from burning more coal if the United States now consumes a quarter of the world's energy? Why should they conserve their biological diversity if the profits of that diversity are monopolized by Northern corporations? Why should they reduce their population growth while stable North American and European populations consume the overwhelming lion's share of what the earth produces and create most of the waste and pollution? If there is to be a new social contract among the nations of the world to alleviate the threat of environmental catastrophe, it would have to entail a more equitable distribution of the earth's bounty. That, at least, was the view from the South.

The view was not widely shared in the North, certainly not in the United States during the administrations of Ronald Reagan and George Bush, and was hardly embraced in Western Europe or Japan. Some environmentalists and a handful of national politicians did take up the theme of unfair and destructive consumption. In his book *Earth in the Balance,* published in 1992, Senator Al Gore called for a "global Marshall plan" to address the environment/development crisis and warned that the rich nations would have to lead by example by conquering their "addiction" to an ideology of consumption.[34] A few northern European government officials joined the call for limitations on the excesses of affluence, but the issue was shunned by most politicians, who found that the idea of an "era of limits" had little resonance among their constituents.

Nor was there much enthusiasm for the idea of sharing sovereignty to save the world. At a meeting in the Hague in 1989, the prime ministers of France, Norway, and Holland offered a plan to create a global environmental legislative body to draw up global regulations and impose sanctions on those who disregarded the regulations. But the proposal failed because of the unwillingness of most governments to give other nations binding powers to restrict freedom to act in what they perceived as their own interests. And, it was suspected, even those nations that sponsored the plan did so in full confidence they would never have to carry it out.[35]

. . .

But the new realities of global interdependence were eroding rigid concepts of national rights and powers. The Helsinki declaration on human rights and the agreements between the United States and the Soviet Union to permit on-ground inspections of nuclear sites had already punched gaping holes in the sovereign armor of nations. Now "a new age of environmental diplomacy"[36] was beginning to build up a body of international law that was progressively converting the notion of sovereignty into a relativistic concept. Nations are increasingly linked—and constrained—by an expanding network of environmental law covering the land, the oceans, island waterways, the atmosphere, and the animals and plants that inhabit the earth.

"International environmental law," noted Nicholas A. Robinson, co-director of the Center for Environmental Legal Studies at Pace University School of Law, "represents a sharp departure from the classic model of diplomatic law, in which a few agreed-upon norms guide how sovereign independent states act to establish their own rules and provide for their own national security. This new legal paradigm, rather than being based solely on the shared needs of sovereign nations, depends on rules that are based on sound ecology and environmental science. . . . In this respect, at least, the 'laws of nature' exist quite apart from the will of sovereign nations." Emerging environmental law, Robinson believes, is likely to play a significant role in shaping the discourse among nations in the coming years.[37]

In some areas of the world, most notably among the nations of the European Union, environmental law is proving to be a rallying point for increasingly cooperative and integrative legal relationships. Margaret Brusasco-MacKenzie, director of international affairs for the European Commission's environmental branch, noted in 1991 that the union had some 250 pieces of legally binding environmental legislation on its books on issues raging from the transportation of hazardous wastes to the protection of migratory birds such as the stork.[38]

Unfortunately, she added, the union's environmental regime is not yet working very well because there are no reliable mechanisms for enforcing the law. Countries in violation can be brought before the European Court of Justice, but its rulings can be carried out only by the voluntary acquiescence of the violating governments.

The problem of enforcement, of course, carries over virtually to all environmental treaties. While the principle of *pacta sunt servanda*, which requires that nations adhere to negotiated agreements, is central to international law, the principle is violated with impunity when it is in the interest of national governments to do so.[39] As international legal scholar Peter Sand pointed out, there are no supranational institutions to approve or disapprove of national activities, nor is there any compulsory jurisdiction to settle disputes that may arise out of an environmental pact.[40] A treaty such as the Convention on International Trade in Endangered Species of Flora and Fauna, for example, is widely flouted because many national governments, particularly in Africa and Asia, turn a blind eye to violations.

There are signs and portents, however, suggesting that, as the mutual dependence of nations for environmental protection and economic progress continues to expand, multinational institutions, particularly the United Nations, will play a more active and direct role in international governance. One such sign is the greatly enlarged peacekeeping function undertaken by the UN in the wake of the Cold War. Eventually there may be UN environmental police to enforce international environmental law. In the meantime, however, the most effective enforcement mechanisms are the activism of domestic environmental organizations, putting pressure on their governments to meet their obligations, and prodding by the news media and citizenry to hold governments accountable for their actions or inactions.

But as with all international agreements, the only guarantee of compliance with environmental law is and is likely to remain the willingness of the nation-state to share voluntarily its sovereignty with the community of nations. As Choucri and North underscore, "Not only is the 'sovereign' state a reality of international life; it is an essential one: We do not know how to manage and regulate the activities of individuals in the absence of . . . 'sovereign' states."[41]

. . .

As East-West tensions ebbed, trade issues began to assume a place of growing importance on the agenda of international diplomacy.[42] The thrust of international trade negotiations since World War II has been to liberalize rules governing the movement of goods and services among nations. Insofar as environmental protection was considered at all in trade negotiations, it was treated chiefly as an impediment to international commerce. The Mexican challenge within the General Agreement on Tariffs and Trade (GATT) to American rules barring the imports of tuna caught along with dolphins was symptomatic, although as opponents of the U.S. tuna rule pointed out, it served the interests of American tuna fishermen as well as of conservationists.

Trade can, however, have a substantial harmful impact on ecological systems and human health, particularly in the poorer countries. Japan's seemingly insatiable need for imported timber, despite having a higher percentage of its territory covered with trees than any other industrial nation, is denuding the forests of Southeast Asia, not to mention the Northwest of the United States, including Alaska. The export of pesticides and hazardous wastes has been blamed for health problems throughout much of the developing world. The havoc wreaked on the fisheries and residents of beautiful Prince William Sound in Alaska when the *Exxon Valdez* plowed into a reef is but one example of the broad environmental damage done by the long-distance international oil market. Aging manufacturing technologies shipped from the industrialized North have created serious air and water pollution problems in the urban areas of many developing countries.

In the past the environmental impact of trade was ignored in international negotiations. A GATT working party on the environment was formed in 1971—but it never met. On the other hand, trade negotiators sought to "harmonize" national environmental standards to the lowest common denominator as a means of facilitating trade. Some developing countries believed that environmental standards were a means used by the United States and Europe to protect their markets from cheaper imports. The importance of commercial trade clearly subsumed the importance of environmental protection.

But this aspect of international relations, too, seemed to be changing in the last years of the century. In February 1992 delegates from around the world descended on Cartagena de Indias in Colombia for the eighth session of the UN Conference on Trade and Development. Previous sessions of the conference, which opened in 1964, had largely served as a forum for the developing countries to air their grievances about the economic inequities imposed on them by the rich industrialized nations.

Cartagena was different, at least rhetorically. There was wide agreement that "the new realities" facing the world called for commitment to multilateralism in the interests of rich and poor alike.[43] "Over the past few years," the final declaration of the conference stated, "the imperative of multilateralism has been intensified by, among other developments, the rise in the level of ecological concern. . . . Transboundary problems such as environmental degradation, fast-spreading epidemics, the consumption and trafficking of illicit drugs, terrorism, migratory movements and the disposal of hazardous waste reinforce this imperative."[44] The statement also expressed concern about the impact that production and consumption patterns were having on the global environment and called for actions by all nations to promote "ecologically sound and sustainable development."[45]

The most recent multilateral trade talks, the Uruguay Round of GATT negotiations, ended in 1993 with some gestures, such as the establishment of an Environment and Trade Committee, that reflect the growing importance of environmental concerns in international relations but with no major changes in its traditional approach to those issues. It was widely believed, however, that the first set of negotiations under the new World Trade Organization, as the GATT system will henceforward be called, would be a "green round."

. . .

It is no accident that the flowering of green diplomacy coincided with the dramatic emergence or reemergence of democratic regimes in central and eastern Europe, Asia, Latin America, and increasingly, in Africa. Experience shows that governments will move beyond traditional and narrow self-interest to pursue the broader good of global environmental and economic sustainability only when pressured by their own constituents to do so; citizens of repressive, authoritarian regimes cannot exercise that pressure. As international policy scholar Caroline Thomas noted,

> The integration of economic and environmental concerns challenges entrenched political and economic interests at all levels. To stand any chance of success, such efforts will have to promote understanding and empowerment at the grassroots level in Northern and Southern countries. . . . Democratization is necessary from the level of the village up to the running of international institutions if the concerns of environment and economy are to be successfully integrated and the cause of sustainable development thus furthered.[46]

Increasingly, people around the world view ecological and economic security as inalienable rights. In its statement "Global Environmental Democracy," the Centre for Science and Environment declared that "the Right to Survival—with a certain modicum of dignity—is the most fundamental of all human rights" and asserted that "all governments must provide their citizens with a justiciable right to a clean and healthy environment."[47] The centre represents an environmentalist perspective from India, but that there is a universal right to a safe, healthy, and pleasing environment is increasingly an article of faith among ordinary people as well as environmentalists around the world.

Not only was an aroused, concerned public putting pressure on governments to act on sustainable development, the very process of environmental diplomacy was becoming discernibly more democratic. An instructive example was the effort to come to grips with the growing threat to the ozone shield. In 1985 governments negotiated the Vienna Convention for the Protection of the Ozone layer. The negotiations were conducted employing the conventional model of "realist" decision making, under which governments alone, acting in what they perceive to be national self-interest, come together to generate an international consensus.[48] The product of that process was a treaty that recognized the ozone peril, created a "general obligation for nations to take 'appropriate measures' to protect the ozone layer," and set up procedures for monitoring the atmosphere. But it did not actually require signatories to do anything to protect the ozone layer. In fact, it did not even name the chemicals that were destroying ozone molecules.[49]

The Montreal Protocol to the Vienna Convention, in contrast, was negotiated under a "polycentric" model of decision making. Not only government diplomats but environmentalists, scientists, corporate executives, and other outside interests, including the media, were integral parts of the process, pressing their own points of view.[50] It was, in effect, a more open, democratized diplomacy, vastly different from the diplomacy of traditional *realpolitik*. The result, as noted at the beginning of this chapter, was a historic agreement by the nations of the world to take sweeping actions for the common good of present and future generations at some cost to current economic well-being. As Ambassador Benedick recognized, "The *power of knowledge and of public opinion* was a formidable factor in the achievement at Montreal. A well-informed public was the prerequisite to mobilizing the political will of governments and to weakening industry's resolve to defend the chemicals" (emphasis in original).[51]

While democracy is essential to environmental diplomacy, the absence of democracy and the suppression of human rights present a difficult barrier to effective international action. James Lilley, a former U.S. ambassador to Beijing who gave sanctuary in the embassy to dissident Fang Lizhi, had this to say in his introduction to Fang's book:

> Right now humanity increasingly faces problems of a global nature: population, energy, environment, atmospheric warming, deforestation and so on. But . . . it is hard to imagine that there could be the necessary understanding and cooperation to solve global problems.[52]

. . .

By 1990, UN Secretary-General Javier Pérez de Cuéllar could state that "it is generally agreed that the environment has moved to the world's political agenda."[53] Lynton Caldwell speculated that global environmentalism might represent the response to a "climacteric" in human affairs, with significant political implications.[54] Gwyn Prins regards "the discovery of planetary systems and pressure for action on the environment" as one of the indications that we are beginning a new "moment" in history. This new era, he wrote, is characterized by a "Gaian" view of the planet, a reference to James Lovelock's Gaia theory, named after the goddess of the Earth, which defines the earth as a single, self-regulating organism. From this perspective, he adds, ecological stress is a "central global political issue."[55]

Moving into the last decade of the century, governments, scholars, and ordinary citizens unquestionably were awakening to the grave dangers to the global environment and to their economic causes and consequences. International diplomacy, which had dealt offhandedly with issues affecting the global commons, began to grapple with these issues in great earnest. International institutions and the international legal system began to widen to accommodate the new understandings about the environment and its relationship with economics and equity within and among nations and between generations.

In terms of what needed to be done to avert the dangers, however, the new geopolitics barely scratched the service. Still unaddressed were the looming problem of climate change, the loss of tropical forests and of biodiversity, the exponential growth of population. While the links between environment and development, between the welfare of the poor countries and that of the rich were now seemingly grasped by policymakers, little was done to alleviate poverty and accelerate economic progress in the Third World. The diplomatic agenda was instead filled with the urgent business created by the collapse of the Soviet Union, the renewed tribal strife in Africa and the Balkans, and a war in the Persian Gulf; with new opportunities for peace in the Middle East; and with shoring up money and equities markets in the industrialized North. Meanwhile, the pigeonholes of foreign ministries around the world were crammed to overflowing with the unfinished business of sustainable development. As Hillary French of the Worldwatch Institute noted, "The problems are growing worse at a rate that far exceeds the pace of diplomacy."[56]

. . .

On the horizon, however, an opportunity for a major breakthrough had arisen. On March 22, 1990, the UN General Assembly, expressing itself as "deeply concerned by the continuing degradation of the state of the environment and the serious degradation of global life-support systems" and as "gravely concerned" that the causes of decline were unsustainable levels of consumption in the industrialized countries and environmental problems caused by inadequate development in the developing countries, passed Resolution 44/228. It called for the convening of a United Nations Conference on Environment and Development in June 1992.

The resolution called for the "highest possible level of participation" at the conference. It would be held in Brazil, in Rio de Janeiro.

A young Maurice Strong presided over the United Nations Conference on Human Environment in Stockholm in 1972. **Photograph courtesy of United Nations/Y. Nagata/ARA.**

In Stockholm, India's Indira Ghandi linked environment and development, warning that "poverty is the greatest polluter." **Photograph courtesy of United Nations/JMcG.**

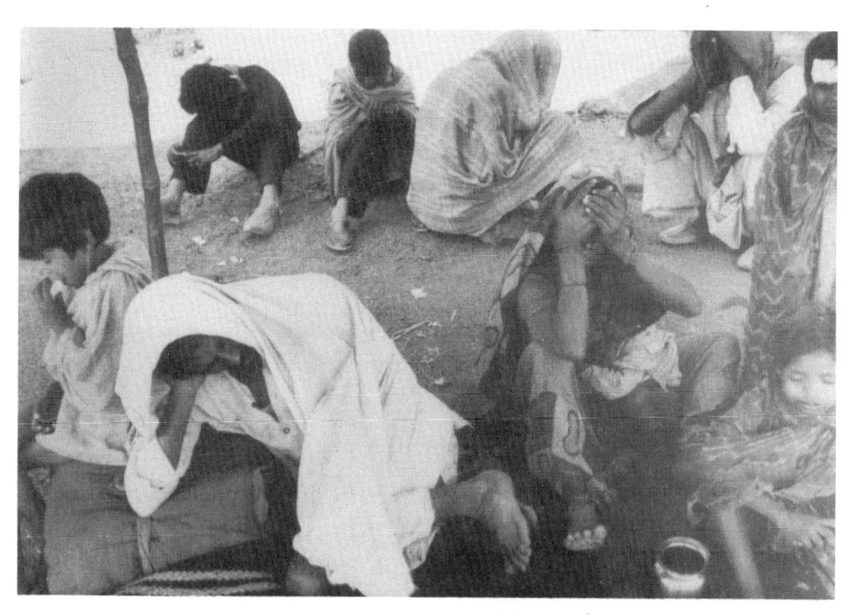

Some of the victims of the leak of poisonous methyl isocyanate from the Union Carbide plant in Bhopal, India. Over 2,000 people died and many thousands more sickened, most of them among the poor. **Photograph courtesy of AP/Wide World Photos.**

Damaged reactor and buildings at the Chernobyl nuclear power station in Chernobyl, USSR, in 1986. The accident spewed radioactive contamination over much of Europe. **Photograph courtesy of AP/Wide World Photos.**

Fishing boats moored in waters of Prince William Sound after it was fouled by a massive spill from the *Exxon Valdez*. **Photograph courtesy of Terrence McCarthy, NYT Pictures.**

Historic photograph shows over ninety of the heads of state and government who attended the Earth Summit. President Bush is in the first row near the middle.
UN Photo 180109/M. Tzovaras.

Three who carried the load of the Rio process. From left: Nitin Desai, Deputy Secretary General; Maurice Strong, Secretary General; Tommy Koh, Chairman of the Preparatory and Main Committees. **UN Photo 177983.**

Norway's Prime Minister Gro Harlem Brundtland addresses Rio conference. Her commission inspired the meeting. **UN Photo 180189/M. Tzovaras.**

Fidel Castro astonished Rio meeting with brevity of his speech. **UN Photo 180076/M. Tzovaras.**

Senator Al Gore addressed Earth Summit shortly before he was nominated as vice president. **UN Photo 180487/M. Tzovaras.**

These two women were among the leaders of the American environmental groups participating in the Earth Summit process. Left: Barbara Bramble of the National Wildlife Federation. Right: Elizabeth Barratt-Brown of the Natural Resources Defense Council.

Alternative summit held by NGOs in Rio's Flamengo Park combined serious purpose with light-hearted fun. **UN Photo 180500/R. Mera.**

Amicable pose of Strong and UN Secretary-General Boutros Boutros-Ghali belies their heated clash in Rio. **UN Photo 180054/M. Tzovaras.**

Worker in the Communa Rio Santiago Cayapas, Ecuador (left), cracking tagua nuts for making buttons. **Photograph © Karen Ziffer.**

Buryat farmers (right) in the Olkhon region of Siberia making hay. The E. F. Schumacher Foundation in the United States is helping the Buryats develop a sustainable, nonpolluting agriculture. **Photograph courtesy of Jachi Nga-Lin Shiu.**

Yolanda Rivera of the Banana Kelly Community Improvement Association and Allen Hershkowitz of the Natural Resources Defense Council (below), standing in the Harlem Rail Yard in New York City, the site of their joint venture, the Bronx Community Paper Company. **Photograph © Harvey Wang.**

Chapter 9

Slouching toward Rio

A certain amount of chaos is O.K. God used it for creation.
—Maurice Strong

The UN Conference on Environment and Development, Maurice Strong claimed on several occasions before it occurred, would arguably be the most important international meeting ever convened.

That was, of course, a rather sweeping assertion. Past international gatherings have altered the world substantially for better or for worse. In this century the Yalta conference shaped the face of Europe for half a century, and the San Francisco conference created the UN system. In the past the Congress of Vienna and the Versailles conference fixed the course of global politics for decades.

Still, Strong's hyperbole was understandable. He had been appointed secretary-general of the conference, the same position he had held in Stockholm. But this conference would be of a different order of magnitude entirely. At the urging of, among others, the Soviet Union,[1] which had boycotted Stockholm, the General Assembly had determined to make it a meeting of heads of state and government—a summit meeting of all the nations in the world. Its mandate was sweeping: to create a new system of global cooperation that would promote economic development and attack poverty in ways that would preserve and protect the earth's vital ecological systems. Unlike the 1972 meeting, the nations of the world would be asked to act not only to protect the environment but at the same time to redress the inequities in the ways that the bounty of the planet was shared by its human inhabitants. It was to address not only the specific problems of environment and development but the *relationship* between environment and development, development and environment.

To fulfill that mandate the conference would place a staggering array of issues on its agenda. It would have to prepare recommendations for the conservation and prudent use of natural resources such as land, water, energy, forests, plants, and animals. It would have to recommend ways to protect global systems, including the atmosphere and the oceans. It would have to address pollution, toxic chemicals, hazardous and radioactive wastes, and other threats to human health. It would have to consider patterns of economic assistance, trade, technology transfers, demographics, poverty, consumption, and the rapid urbanization of the human community. It would have to make links between all of these issues in ways that could be translated into policy by governments and international institutions. Perhaps most difficult, it would have to seek ways to finance all of the actions it would recommend.

As preparations for the conference progressed, it became appallingly evident that there was hardly an issue under the sun that did not somehow fall under the rubric of environment and development.

The significance of the conference, however, would be much more than the sum of its many parts. By an accident of history the Rio meeting—the Earth Summit, as it soon came to be called—was to take place in a time of phenomenal flux and fluidity in the global body politic. In particular, the disintegration of the Soviet empire, the evaporation of the Communist alternative to market economies, and the end to the Soviet-American struggle for global ascendancy had canceled many of the rules by which the complicated, hazardous game of international politics had been played for nearly half a century. New rules would now have to be framed by the community of nations.

The Rio conference thus would be the first post–Cold War test of whether the global community yet possessed the will and the wisdom to create a new system of collective security based not on fear of mutual destruction but on recognition of the urgent necessity of cooperation to preserve the planet and improve the human condition. It would be the first major opportunity for the UN system to demonstrate that, freed of the shackles imposed by the long East-West confrontation, it could be an effective force for global peace, justice, and progress. It would be the first significant showdown between the North and South since the geopolitical fault line rotated away from the half-century East-West divide, offering an opportunity to defuse tensions between the affluent and poor countries—or to ignite them.

The Earth Summit also would be watched as an indicator of the roles that nations and blocs of nations would assume in the new era of international diplomacy. Would the United States use its now unchallenged position as the most powerful nation on earth to exercise global leadership, and if so, what kind of leadership? What role would be taken by Japan? The European Union? Where did the nations of the splintered Soviet bloc fit in the new diplomatic puzzle? How would the nonaligned nations readjust now that the bipolar alignment of world politics had disintegrated?

"This is," said Maurice Strong at the beginning of the process, "the first time in history that the leaders of all the countries of the earth will assemble to make decisions that will literally shape the future of our world."[2]

. . .

After stepping down as director of the UN Environment Program (UNEP), Strong had returned to private business, serving as chairman of his own firm, Strovest Holdings in Canada, and as chairman of the Canada Development Investment Corporation. But he remained active—the record suggests hyperactive would be a better word—on international issues. He headed the UN's famine relief operations for Africa in 1985–86. He served as chairman of the International Union for the Conservation of Nature and Natural Resources, president of the World Federation of United Nations Associations, and as an active member of the Brundtland Commission.[3] He also was secretly asked by UN secretary-general Javier Pérez de Cuéllar to negotiate the arrangements for peace that were to follow the cease-fire in the war between Iran and Iraq.

Sweden had wanted to play host again to a UN conference on the environment, which was to be held on the twentieth anniversary of the Stockholm meeting, and the Swedish government pressed for a Swede to be named chairman of the preparatory committee on the new conference. Canada also wanted to hold the meeting, but when Brazil, as a representative of the developing nations, asked to hold the meeting in Rio, both Canada and Sweden stepped aside. Meanwhile, unknown to most people, Pérez de Cuéllar had already made it clear to Strong that he would like to appoint him to head the the Earth Summit secretariat.

Although a rich businessman from an industrialized country and opposed by some Third World activists and women's groups as a representative of the same old, white male, wealthy, northern established order, Strong had won the trust of most of the developing countries by his many efforts on their behalf. Most environmentalists were enthusiastic about his appointment, although a few were troubled by his management of his huge land holdings in Colorado. Canada's prime minister at the time, Brian Mulroney, nominated Strong to the position, even though he was affiliated with an opposing political party. Although he had not sought the secretary-generalship, Strong agreed to accept it.[4]

The first organizational meeting for the conference was held at UN headquarters in New York in March 1990. From the outset, Strong faced a task so formidable and daunting that a less confident man might have quailed before it. But he possessed an optimism so powerful that it could be mistaken for arrogance—or naïveté—and he plunged cheerfully into the labyrinth.

His initial resources were modest. For a time he was a one-man band, doing his own recruiting, setting up schedules, prosyletizing to governments, searching for other resources. He was given a small budget and had recruited a tiny core staff. Eventually a private organization called EcoFund '92, created by Benjamin Read, a dedicated American public servant and humanitarian, raised substantial funds, enabling Strong to enlarge his secretariat and engage in a variety of activities that would not have been possible given the frugal resources provided to him by the General Assembly. Although allotted only twenty-two positions for his secretariat, far too few for the staggering work load he would face, Strong eventually was able to enlist the services of some seventy highly qualified professionals from every part of the globe, as well as obtaining volunteer assistance from a much larger network of advisors and helpers from the worlds of diplomacy, finance, and academe.[5]

The preparatory committee for the Rio meeting consisted of all member nations of the UN General Assembly, much too unwieldy a body to organize the details of the conference, to draft the initial negotiating documents, and to prepare the political structure needed to assure the participation and cooperation of governments. That burden would fall on Strong and his secretariat and on three "working groups" appointed by the member nations.

The arrangement had some beneficial aspects. For one thing, the participation of all governments in the preparations lent the process significant weight. It gave governments a stake in trying to reach a successful outcome. But it also had substantial drawbacks. As Strong noted, "There is always tension between secretariats and governments. Governments always have a natural inclination to be somewhat frustrated by their dependence on secretariats. They want to make sure they keep secretariats in their place. Initially, there was no question that they did not want the thing secretariat driven. But I knew that if we didn't do our homework, nothing would get done. So I was always very deferential to governments but, nevertheless, I kept putting our proposals forward."[6] Strong also had to win the confidence of the various UN departments and agencies, independent fiefdoms jealous of their prerogatives and, at first, suspicious of the entire Earth Summit process.

One of the most difficult tasks Strong faced at the outset was translating the General Assembly resolution authorizing the conference into a workable agenda. Resolution 44/228 contained very specific language, much of it adopted at the insistence of the Group of 77, the caucus of developing countries (actually, the number of members had grown to more than 120 countries since its founding), calling for such things as additional financial resources, the transfer of technologies on favorable terms, and changes in patterns of production and consumption. Such programmatic demands, Strong felt, would have preempted the conference and made meaningful negotiations difficult if not impossible. "When I looked at it I said, 'You can't make a conference out of this.'"[7]

Instead, Strong conceived of a conference that would revolve around six themes or products. One would be an "Earth Charter," a kind of *Terra Carta* that would lay out broad principles for preserving the planet and enhancing the welfare of the humans who inhabit it. The second would be an "Agenda 21," a detailed action program for carrying out the principles and for achieving a series of goals pointing toward environmentally sustainable economic development in the twenty-first century. Other elements for negotiation at the conference would be the financial means for carrying out Agenda 21, technology transfers, and institutions to follow through on the decisions of the conference. Finally two treaties—one dealing with climate change, the other with preserving the planet's biological diversity—would be negotiated on separate but parallel tracks with the conference and signed in Rio.

Whether or not Rio would be the most important conference in history, it certainly would be one of the most difficult and complex. Strong, his secretariat, the UN agencies, the governments of the world, and the citizens groups and other private organizations that wanted to leave their mark on the summit had just two years to get ready.

. . .

The first of four preparatory committee meetings—"prepcomms," as they were universally called—was held in Nairobi in August 1990. It did not go particularly well. Government delegations were largely ill-prepared; many did not fully grasp what the ultimate goal of the conference was to be. The prepcomm was held immediately after a meeting of the UNEP governing council, and many of the delegates were holdovers from that gathering. There was mutual incomprehension between the regular UN diplomatic corps, which comprised the bulk of the delegations, and the few experts on economics and science sent out by their capitals. Strong recalled a day when two visitors came separately to his office in Nairobi, one "an expert who said 'all this damn political talk about process is so frustrating' and another, regular UN delegate who said, 'You know, all these experts are really a nuisance. They complicate our life. They know issues but don't know anything about the UN.' "[8]

Governments testing the unfamiliar geopolitical waters were reluctant to commit to substantive agreements on the content of the conference and even had difficulty concurring on process. At the outset they were also intent on staking out their maximum demands. This was particularly true of the developing nations, which saw the conference as an opportunity to exact concessions from the North in the form of increased transfers of money and technology. One member of the secretariat staff noted that because of the need for their cooperation in protecting the global commons, "the South is no longer a supplicant. Of course, the problem is how to take that and make it a workable political package."[9] But the industrialized countries, particularly the United States, which made it clear from the outset that it would not agree to the "new and additional" sources of funding demanded by the Third World countries to protect the environment, also took hard-line positions.

Nevertheless, the first prepcomm was able to keep the conference process moving forward, if painfully, by "avoiding paralyzing ideological positioning" and by establishing a process of negotiating by consensus.[10] Delegates also agreed to form two working groups to deal with specific environmental issues. Group I, under the chairmanship of Ambassador Bo Kjellen of Sweden, would deal with biodiversity, biotechnology, desertification and soil erosion, and atmosphere and climate. Group II, chaired by Ambassador Bukar Shaib of Nigeria, would work on issues related to protecting the oceans and freshwater resources and hazardous wastes. The prepcomm requested all governments to prepare national reports for the summit, describing their own experiences with environment and development.

And in an action that greatly enhanced the prospects for ultimate success at the summit, delegates to the Nairobi meeting elected Ambassador Tommy T. B. Koh of Singapore as chairman of their deliberations, a position he would hold until the Rio Conference, where he served as chairman of the Main Committee. Koh, in turn, appointed a substantial number of vice chairmen to preside over informal working groups during the preparatory process.[11] A bright, tough, but amiable and witty lawyer and legal scholar, Koh had been one of the youngest diplomats ever to head a national delegation to the UN. With a singular talent for mediation and compromise, he had shepherded the tortuous Law of the Sea negotiations through their conclusive phases. All his skills would be needed in the Earth Summit process.

One of the most significant decisions taken by the first preparatory committee meeting and later endorsed by the General Assembly to cover the entire conference process, including the summit, was to agree to include nongovernmental organizations (NGOs) as an integral part of the process. The agreement to integrate the NGOs was not easily reached; there was substantial opposition, particularly among some of the developing countries with little experience in participatory democracy. It required painstaking negotiation and wheedling by Strong and Koh to obtain their grudging acceptance.[12]

Once let in the door, however, nongovernmental actors, such as environmental, human rights, women's, youth, indigenous peoples', business, and labor groups among others, were active and insistent contributors to the process. While they were not actually permitted to take part in the formal political negotiations, they were amply able to contribute to them—to submit papers, speak at plenary sessions, lobby in the corridors and committee rooms, and even be part of some national delegations. They added energy, enthusiasm, fresh ideas, and a critical perspective, often a firsthand perspective, on what was actually happening in cities, villages, and forests. They also infused an occasional touch of radicalism into the deliberations.

. . .

The acceptance of the NGOs by governments remained tenuous throughout the process. Barbara Bramble, who oversees international activities for the National Wildlife Federation, later recalled that Strong was continually begging them: "Please don't blow up the process and please give me some ammunition to show that what you are bringing is of use." One factor that helped was that there were relatively few NGO groups in Nairobi, perhaps forty, a number with which the national delegations and conference secretariat were able to deal with relative ease. If governments knew that the number would mount into the thousands by the time the Rio conference convened, they might have hesitated to agree to allow them such direct involvement.

The Nairobi prepcomm was valuable as an "educative process," as Strong put it, setting the broad range of issues before government delegations, exposing to a harsh but necessary light the political difficulties the conference would face in dealing with those issues, and uncovering some of the institutional shortcomings of the UN system. It also gave a first glimpse of the sometimes maddening bureaucratic and even clerical procedures that would have to be overcome if the Summit was to succeed.

There were also the first glimmerings of comprehension among national delegations that the problems of environment and development were, in fact, linked with the interests of the rich and poor nations. But by the end of that first prepcomm it was questionable whether the Earth Summit would even take place, much less succeed. The delegates did not approve any of Strong's six themes for the summit, not even the Earth Charter. They did not authorize the creation of a third working group to deal with "cross-cutting" issues such as financing, technology transfer, and legal and institutional arrangements. Some governments were suspicious of Strong's secretariat, fearing that it was leading them in directions they did not want to go—for example, protecting their forests for the common goal of combatting global warming or, in the case of the United States and some other rich nations, coercing them into financial concessions they did not want to make. At the end of the Nairobi meeting the secretariat had no mandate to offer substantive proposals. Governments were also wary of the motives of other governments, particularly along the North-South divide, fearing the other side was looking only to its own interests.

The Nairobi prepcomm, Strong recalled after the Rio meeting, was one of the low points of the whole exercise. "It was a very unruly, divisive, frustrating operation. Nobody at that meeting could have believed we would be able to negotiate a document like Agenda 21 by the end of the process."[13]

. . .

The second prepcomm, held in Geneva's Palais des Nations in early spring 1991, was hardly less contentious. The sharply diverging North-South views of what the conference should attempt emerged in harsh relief. The issues related to the economic needs of the developing nations, and their demands on the industrialized countries burst into an often inflamed debate that did not cool up to the last gavel in Rio.

Statements by the chairman of the Group of 77 and the leader of the U.S. delegation offered an early but painfully clear demonstration of the core issues at the conference and the high hurdles the preparatory committee would have to leap to resolve them. They deserve quoting at some length.

Ambassador Kofi Awoonor of Ghana, speaking for the 77, summed up the position in particularly blunt terms:

> The fundamental issue of global economic imbalance remains the principle cause of third world poverty, which in turn promotes environmental degradation. . . . It is our firm belief that the untenable situation in which the industrialized countries virtually confiscate our products by paying prices nowhere near parity in relation to the products they sell to us should not, in good conscience, be allowed to persist a day longer. When . . . we in the poor nations pay for the unsustainably high standard of living in the developed nations . . . the very development assistance which comes to us arrives with paternalistic and humiliating conditions; or when basic development technology arrives with price tags that deepen not only our poverty but extends our condition of peonage to our so-called benefactors; we think the time has come we must resume the North-South dialogue on these serious matters. . . . Until this situation is redressed by a serious, agreed and equitable global economic program, we shall insist on compensatory financial outlays from the industrialized countries, which no one should construe as charity but necessary transfers for what is due us for years of continuous labor . . . it cannot be expected that, because of the present perverse economic order, those who earn $200 per capita in the great democratic free-market place, are the ones to make sacrifices so that those who, by dint of the massive advantages of technology and an exploitive international economic regime, earn $10,000 per capita can breathe cleaner air or escape the tormenting discomforts that global warming may bring in its wake.[14]

Shortly thereafter, the head of the U.S. delegation, Assistant Secretary of State E. U. Curtis (Buff) Bohlen, took the floor at the plenary session to outline the American position and essentially dumped cold water over the Group of 77 demands. The word "compensation," he said, was "not particularly useful." The United States was already doing its share for environment and development, having given some $250 billion in foreign aid over the past twenty years and was currently spending $150 million a year on aid programs specifically aimed at environmental problems. If more money was to be forthcoming for environmental goals, Bohlen said, it would have to come by redirecting "scarce resources" to countries that had "demonstrated their capacity to use those resources by adhering to environmental agreements." The only other sources of new funding would have to come from "private sources," he added, meaning investments by private banks and corporations.[15]

The line was clearly drawn in the sand. The next day, at a regular morning meeting between Koh and Strong and their staffs, Koh said: "I was very disappointed in the debate. It takes me back fifteen years."

The developing countries took and held the position that the rich nations, through their excessive consumption and destructive technology, had caused the global environmental problems and that they, the poor, should not be asked to make additional sacrifices to solve those problems. If the developing countries were to cooperate in programs to safeguard the environment, they would have to be given new and additional financial and technological resources to do so. And the industrialized nations would have to reduce their consumption and production of waste.

Among the industrialized countries, the United States adopted the hardest line in opposition to the demands of the South. But while a number of the rich nations gave some rhetorical support and vague economic promises to the developing countries, it was an open secret among the delegations that the countries of the European Union, with a few exceptions, were pleased to have the Bush administration serve as the naysayer for all of them. For most of the presummit process, the U.S. position was adamant: No concessions on new and additional resources, no transfer of technology on concessional terms, no concessions on consumption by the rich, no new economic order.

The United States also dug in its heels on other key issues. In the parallel negotiations it refused to agree to specific targets and timetables in the climate treaty and played hardball in negotiations on the biodiversity treaty. It tried to keep the preparatory committee from considering environmental problems in Antarctica and to block any reference to the impact of military activities on the global ecology. Wanting to keep open the option of disposing of its decommissioned nuclear submarines at sea, the U.S. delegation sought to bar any agreement that prohibited dumping of radioactive materials in the oceans. President Bush long declined to say whether or not he would attend the summit meeting in Rio.

As Jessica Tuchman Mathews, then with the World Resources Institute, put it, the U.S. delegation spent much of the summit process in "a defensive crouch. The negotiations are viewed as a job of damage limitation, not as a potential contribution to international security and economic growth."[16]

The posture of the United States during the summit negotiations was a textbook illustration that the *realpolitik* that motivates participants in international negotiations is not necessarily or even usually the interests of their nation. Their positions are frequently driven instead by the narrow and immediate partisan political needs of whoever is in power.

The U.S. delegation included a number of talented and dedicated public servants who personally believed in the need for global cooperation to create a new system for protecting the environment and promoting economic equity.

Buff Bohlen, for example, had been an officer of the World Wildlife Fund before returning to the State Department. President Bush himself had said that he intended to be the "environmental president," and his record before becoming Ronald Reagan's vice president indicated that that is where his instincts lay.

But as the Summit process unfolded, President Bush was preparing for a reelection campaign and feeling pressure from the increasingly dominant right wing of the Republican Party. He also was facing a challenge for his party's nomination from Patrick Buchanan, a right-wing ideologue and intemperate propagandist. In John Sununu, Bush's chief of staff until he was pressed to resign, the right had an aggressive agent strategically placed at the inner core of the White House. Bush, feeling politically vulnerable, believed he had to placate the right of his party, whose marching orders were summarized by the Heritage Foundation, one of its think tanks.

Concessions at the Earth Summit, according to the foundation, "could affect profoundly America's economic growth, productivity, and international competitiveness." It urged, among other things, that the United States oppose any proposals for spending more money on the environment, avoid specific targets for limiting greenhouse gases, and protect private intellectual property rights on technologies sought by Third World countries on concessional terms.[17] Those rules were faithfully observed by U.S. negotiators through the summit process, however reluctantly, at the instruction of the White House.

National and regional politics also affected the negotiating posture of the western European countries. Many expected Europe to play a forceful role in the Rio process. The European Union had sought and received from the UN the right to speak as a separate entity during the proceeding in what was to be the emergence of a unified European diplomatic voice. A number of smaller European countries, including the Netherlands and Scandinavia, were staking out very forthcoming policies on development assistance and cooperation to protect the global environment. Europe's Green political parties seemed to be on an ascending trajectory, and a leader of the French Greens, Brice Lalonde, was his country's environment minister.

As it turned out, however, Europe as Europe did not have as major an impact on the negotiations as they had hoped. The timing was wrong. Engaged in a delicate, difficult effort to strengthen their own political unity through ratification of the new Maastricht Treaty, which would extend and accelerate the integration of many previously sovereign prerogatives of Union members, the governments of western Europe were unwilling to spend too much political capital on the Earth Summit process. Most of the European countries, moreover, were entering into one of the worst of their post–World War II recessions and were preoccupied with their own economic problems, not with issues of global sustainable development.

Japan? As usual in international negotiations not directly related to their own economic interests, the Japanese were diffident and enigmatic. Strong would soon try to change that.

. . .

As the second prepcomm drew to an end, it appeared that the delegates were talking about two different conferences: one about a sustainable environment, the other about economic development. One frustrated member of the U.S. delegation felt that "the international structure to deal with these issues is totally inadequate. You can't negotiate issues as complex as these in a 150-nation forum." And a member of the secretariat complained that "the U.N. system is not geared for the 21st century."

But Strong was not discouraged. "A certain amount of chaos is O.K.," he observed. "God used it for creation." He and Koh had won some key victories. By consensus, the prepcom agreed to consider Strong's six themes and instructed the secretariat to begin drafting the Earth Charter and agenda 21. In effect, governments agreed to consider Strong's program as the basis of the Earth Summit negotiations, although there was no guarantee that those documents would ever be formally accepted. They also agreed to form a third working group to deal with the cross-cutting issues. Strong and others were not at first entirely pleased with the selection of Bedrich Moldan of the Czech Republic as its chairman, unsure that he had the experience and clout needed to preside over the complex and divisive negotiations confronting his group.

The industrialized nations, including the United States, acceded to Third World demands that poverty be placed on the agenda as an obstacle to sustainable development. The NGOs—there were 173 of them at the second prepcomm but only about 20 from developing countries—began to play an active and highly visible role. Women's groups in particular began to make their voices heard. In a meeting between NGOs and Strong, Bella Abzug, the feminist leader and former U.S. Congresswoman, urged that the women's role in sustainable development be acknowledged in a separate chapter of Agenda 21 and, in her acerbic fashion, stated that humans "have done everything in pairs since Noah except participate in government."

What was left by the end of the prepcomm was what Clif Curtis of Greenpeace described as a "complex Rubik cube." Strong, his secretariat, and his allies would have only a few months until the next meeting, in August, to begin solving the puzzle: to prepare the draft documents, reconcile conflicting views, muster political support, push recalcitrant governments into line, and find answers to the myriad questions that governments had put to them.

. . .

While his secretariat labored feverishly at gathering data and preparing documents at its parklike headquarters in Conches, on the outskirts of Geneva,

Strong sped indefatigably from continent to continent, country to country, city to city, on an evangelical mission to convert the world to the cause of the Earth Summit and sustainable development. He visited Washington, D.C., repeatedly to persuade the White House to announce that Bush would go to Rio and in a largely fruitless effort to build enthusiasm for the summit in the Bush administration. He traveled to Third World capitals to reassure their governments that the summit would stress the development side of the sustainable development equation. He spoke at endless conferences, seminars, and press conferences to alert the world public and the business communities to what was at stake in Rio. While not in the best of health, he clearly reveled in his exhausting schedule. "What could be more interesting in life?" he said at the end of one long day. "I thank God every day for the opportunity."[18]

What Strong and the conference needed most was leadership from a major power and a jump start on the process of financing the sustainable development programs.

Seeking both, Strong went to Tokyo.

Received as a world leader, Strong met with the cream of Japan's power elite, including then prime minister Toshiki Kaifu, the top leadership of the powerful Ministry of International Trade and Industry (MITI) and other ministries, the reigning bureaucrats of the dominant Liberal Democratic Party, and the directors and members of Keidandran, the highly influential business and industry association. Using a judicious mixture of flattery and persuasion, Strong pounded home the message that the Earth Summit was an opportunity for Japan, proscribed by its constitution from exerting military power, to use its economic might to exercise world leadership. He also sought specific financial commitments from the Japanese and, above all, a Japanese initiative in setting up a permanent global financial mechanism to fund the programs of Agenda 21.

Meeting with Kaifu in the prime minister's office, a rather small, plain, wood-paneled room decorated with a limp Japanese flag and a picture of the late Emperor Hirohito and his wife, Strong urged him to announce that he would personally attend the summit as well as push for a forceful reference to the Rio meeting at an upcoming meeting of the heads of the major industrialized nations. "We need your leadership in Rio," Strong emphasized. "You have set an enviable domestic environmental record. We know that you have performed better than any industrial country in cleaning the environment and that prepares you for international leadership."

Kaifu's response was friendly but noncommittal: "Japan has always placed a high priority on environmental issues. I am aware of a need for a global approach. We will be working with our cabinet members."[19]

At a breakfast meeting of Keidandran, Strong appealed to the industrialists' self-interest by stating that "no country is better suited than Japan to participate in the econ-industrial revolution. . . . Enlightened business leaders see that environment is driving the new wave of business opportunity." He pointed out that with Japan's technological capacity and with the hundred-year program planned by MITI to develop and market environmental technologies, Japanese industry was well positioned for a global economy based on ecological principles. He then added that a major effort would be needed to mobilize funds for the global environment and suggested that Japan could make an annual contribution equal to what it spent on the Persian Gulf War, $12 billion. In international affairs, he said, "You don't have to wait for the United States. The United States has taken the major military role. Why doesn't Japan take the lead in this nonmilitary capacity?"

Strong's key meeting in Tokyo was with former prime minister Noboru Takeshita, out of government and in private business but, as de facto leader of the most powerful faction of the Liberal Democratic Party, still Japan's most influential politician. Strong wanted Takeshita to use his influence to open Japan's purse strings for global environmental programs, particularly in the developing world. "Some of the people in your faction," he said to the former prime minister, "are urging Japan to set aside 1 percent of its GNP for peaceful initiatives in the environment. If this could be Japan's initiative, it would put Japan in the forefront of the new push for global security." But above all, Strong wanted Takeshita to call and be chairman of a meeting of "eminent persons," to be held just before the summit, that would plan ways to finance sustainable development.

Takeshita said somewhat coyly he was not qualified to head such a meeting but then quickly indicated that he would be honored if Strong offered him the chairmanship. When would it be and who would attend, he wanted to know before agreeing to accept the invitation.

It would be in the spring of 1992, and attendees would be "very influential people" such as former U.K. prime minister Margaret Thatcher; Paul Volcker, former chairman of the U.S. Federal Reserve Board; and Stephan Schmidheiny, the billionaire Swiss businessman and chairman of the Business Council for Sustainable Development, Strong replied.

After a pause, Takeshita said: "I would like to think seriously about the timing."

Although he did not receive any firm commitments, Strong was pleased with the results of his trip to Tokyo. He was fairly certain that he had persuaded the Japanese to be forthcoming on the financial issue and to take a leadership role in Rio. And he rightly guessed that Takeshita would agree to host the finance meeting.

By the end of the week, however, he was weary and, for one of the few times during the process, uncharacteristically down. "I can be very pessimistic on an intellectual level," he confided during a quiet conversation. "Environmental problems are like a cancer spreading insidiously through the body. They will probably kill us eventually, but the symptoms are not acute enough to prod us into saving ourselves."[20]

. . .

The third and penultimate preparatory committee meeting, also held in Geneva, was opened by UN secretary-general Javier Pérez de Cuéllar, who urged the delegates to find "a broad consensus for a more stable, equitable future." The foundation of that consensus, he said, would be a "genuine North-South partnership," massive participation by non-governmental organizations and great determination, particularly by the developed countries, which would have to change their patterns of consumption and provide new resources for the developing countries.

"Summit planet earth," he declared, "will be one of the major steps toward the 21st century."

The speech was something of a valedictory for Pérez de Cuéllar, who would soon complete his term of office and depart. The pending change in leadership, along with mounting demands for reform and restructuring of the UN system, further muddied the already murky political waters through which the summit process was navigating. Strong was among those being mentioned to succeed to the secretary-generalship. So was Prime Minister Gro Harlem Brundtland of Norway, whose report had galvanized the UN to call the environment and development conference.

For the diplomatic community and for the entire world, a shocking and even more disorienting event had taken place just a week before the third prepcomm began. In the Soviet Union a junta of hard-line communists had staged a coup against the reformist regime of Mikhail Gorbachev, placed Gorbachev under house arrest at his vacation retreat in the Crimea, proclaimed a state of emergency, and sent troops and tanks into the streets of Moscow. But led by Boris Yeltsin, the people of Moscow defied the junta, and within two days the coup attempt had failed and Gorbachev was back in Moscow. But he returned to an utterly different political climate and within a few months would resign, and the Soviet Union would disintegrate.

At the Palais des Nations the delegates drifted about looking for familiar moorings. Knots of dark-suited diplomats gathered in the huge delegates' lounge, with its floor-to-ceiling windows overlooking wooded lawns sloping down to Lake Geneva, could talk of little but the dramatic events in the Soviet Union.

Strong quickly placed those events into the context of the work of the prepcomm. "Preparations for the Earth Summit," he said in a speech to the first plenary session, "are taking place at a time of unprecedented change for both the peoples and the governments of the world. . . . The end of the cold war and the restructuring of military budgets make what was once only a dream now a realistic prospect. To forge a new partnership, to translate the dream into a reality is the primary responsibility of this committee."

It was apparent that the government delegations were prepared at last to begin serious negotiations on the substance of the conference, sometimes using the documents prepared by the secretariat as the basis for discussion, sometimes not. By some enigmatic process of political osmosis, the committee had given its tacit assent to the creation and adoption of Agenda 21 and were now engaged in protracted and detailed negotiation over its many elements. As a delegate from Norway noted, Agenda 21 was now understood as "an operational action plan" for achieving sustainable development. It would not be a legally binding document but would possess "a high degree of political commitment."

Virtually every chapter of Agenda 21 became a combat zone as countries and blocs of countries sought to insert their own favored issues and delete any sentence, phrase, or comma deemed contrary to their perceived interests. It also became apparent that the short, inspirational Earth Charter sought by Strong and others would not be attainable. Nation after nation wanted to insert its own pet hobbyhorses into the document, and by the end of the prepcomm more than one hundred countries, from Australia to Zaire, had submitted language. Instead of a charter there would be an intensely negotiated "Rio Declaration on Environment and Development" eventually submitted for approval at the summit.

It also became clear at this meeting that there could be no agreement on a binding treaty to protect the forests of the world. Countries such as Malaysia, the Philippines, and India rejected any commitment to limit economic exploitation of their forests as an intrusion on their sovereignty. Some of their delegates pointedly noted that the rich nations had already exploited their forest resources to the fullest but were now asking the poor countries to refrain from doing the same thing in the name of international cooperation to protect the environment. A number of African countries called forest preservation a northern issue and insisted that a treaty requiring international action to halt the spread of deserts was far more crucial to their needs and to the goal of sustainable development. As a compromise there was an agreement to negotiate "Principles on World Forests," which might or might not become the basis for a binding convention on forest preservation at some future point down the road from Rio.

Governments also were unable to agree on what kind of institutional changes would be needed inside the UN to follow up on the decisions taken in Rio or even if there should be any changes.

The chief issue of contention, however, continued to be money. Nay Htun, a veteran international civil servant from Myanmar and a key member of Strong's secretariat, commented toward the end of the third prepcomm that "there is one and only one issue remaining—new and additional financial resources." The developing countries continued to insist on those new resources to pay for any new activities that would be required for the action programs of Agenda 21. And they also demanded much greater participation in the process of deciding how those resources would be allocated and deployed—no more diktats by World Bank bureaucrats or finance ministries in the industrialized capitals. But the North, particularly the United States, remained unwilling to enter into new economic commitments, all the more so because no one had any idea how much money was involved.

The next and final prepcomm would be held in March at UN headquarters in New York City. The secretariat had six months, not only to come up with the details of Agenda 21 but also to provide governments with estimates of how much it would cost to pay for each item on the agenda.

"I don't need to tell you," Strong told his secretariat at a postmortem of the Geneva meeting, "that we have a lot of work to do."

Chapter 10

To the Wire

We haven't the time to take our time.
—Eugene Ionesco, *Le Roi se meurt*

Grimy snow was still piled on the streets of New York and a bitterly cold wind blew off the East river as delegates gathered at UN headquarters at the beginning of March 1992. But Rio de Janeiro was only three months away. A sense of urgency permeated the fourth and final preparatory committee meeting. So too, at the outset, did an almost palpable mood of gloom. An enormous work load faced the delegations, and the outcome was far from certain. Almost nothing had been agreed on, and the negotiators had only five weeks in which to reach closure on perhaps the most diverse and complex set of issues ever set before an international forum. If problems were sent unresolved to Brazil, there was little likelihood that many of them would be decided by heads of state and government. There was a nagging suspicion among governmental and nongovernmental participants that traditional geopolitics might not be capable of a timely response to environment/development crisis.

It also soon became apparent, however, that national delegations had determined that the time for rhetorical posturing had passed and that they were at last ready to enter into serious negotiations. They were now "sherpas" prepared to lead the way to the summit. Speakers from the developing countries, who had focused almost exclusively on development issues at the early meetings, now began to acknowledge the seriousness of the environmental threats to the human community. Time was growing short, not just to prepare for a diplomatic conference but to confront those threats. Ambassador Jamsheed Marker of Pakistan, who had taken over as chairman of the Group of 77, referred in a speech at an evening plenary session to a recent report by the World Resources Institute; the report had noted that the amount of arable soil lost around the globe due to deforestation and desertification over the previous forty years was equivalent to the entire land mass of India and China combined. "This report should concentrate our minds wonderfully," he said.[1]

. . .

Maurice Strong's secretariat had done its part between prepcomms. It had prepared detailed documentation that would form the basis for what would turn out to be an eight-hundred-page Agenda 21. The document, envisioned as a translation of the concept of sustainable development into systematic, programmatic actions, would be constructed around a series of themes:

- "The Prospering World," the theme of achieving economic growth with sustainability, of integrating environment and development in governmental decision making.
- "The Just World," which could be achieved by combatting poverty, changing consumption patterns, dealing with demographic dynamics, and addressing human health problems.
- "The Habitable World," which addressed issues of sustainable human settlements, including urban water supplies, solid waste management, and urban pollution and health.
- "The Fertile World," which dealt with the efficient use of land, water, and energy resources; sustainable use of forests; management of fragile ecosystems such as mountains, coastal areas, and islands; conservation of biological desertification; and environmentally sound management of biotechnology.
- "The Shared World," an approach to protecting global and regional resources, including the atmosphere, the oceans, and marine life.
- "The Clean World," to be achieved by managing toxic, hazardous, and radioactive waste.
- "The People's World," a world of participation in and responsibility for the sustainable development process by a broad range of groups including, women, children and youth, indigenous people, nongovernmental organizations (NGOs), farmers, local authorities, trade unions, business and industry, and the scientific and technological community.[2]

Finally, the secretariat suggested the means essential to carry out the programs of Agenda 21: information for decision making, national mechanisms and professional capacity for carrying out the programs, access to the science necessary to underpin sustainable development, access to environmentally sound technology, legal instruments and institutions for international cooperation—and, of course, money.

Between prepcomms the secretariat also had accomplished the heroic task of calculating price tags for each of the dozens of programs contained in Agenda 21. The process by which the figures were reached would not stand up to careful scrutiny. A member of the secretariat related how the cost figure for one major item on the agenda was arrived at. "We were at a party and drinking a lot of wine, when somebody said, 'What about this number?' And we all said, 'Yeah, let's take that number.' "

Strong however, defended the cost figures presented by his secretariat.

> We never represented these as firm, engineering-type estimates. We represented them as the best estimates that knowledgeable people who understood the issues can make. There was a lot of intuitive analysis involved, no question about that. We never said otherwise. But as orders-of-magnitude figures they stand up. Some of the industrialized countries were appalled by the size of the numbers and wanted to attack them. But they didn't because they couldn't find any fundamental errors.[3]

The total cost of the Agenda 21 programs would be $600 billion a year, about $125 billion of which would have to be assistance provided to the developing countries by the rich industrialized nations. That amount was $70 million more than the total aid going to developing countries at the time[4] and a sum that seemed especially staggering at a time of serious recession in most of the North. In many of those countries, official development aid had been shrinking, not expanding. Although Strong had made it clear he was not seeking traditional development aid but instead a reorientation of existing of economic priorities through incentives and penalties that would produce more resources for the poorer nations, but the industrialized countries did not arrive in New York ready to create a new economic order. But as Jamsheed Marker, among others, pointed out during the New York meeting, $125 billion was roughly what the recent Persian Gulf War had cost, and the industrialized nations had managed to come up with that money without inflicting any perceptible economic hardship on their citizens. "We refuse to accept that the funds are not available," Marker said. "We do not need a specific figure immediately. We know that it is a difficult time in the North. We do need a commitment that they will do it."[5]

. . .

As the working groups, both formal and informal, settled down to negotiation of Agenda 21 and the Rio Declaration, the truth of the old saw that "the devil is in the details" became quickly apparent. Disputes and deadlocks erupted on virtually every document, every chapter, every paragraph, comma, and period. Reports made by members of the secretariat and conference staffs during their morning meetings in the third week of the prepcomm convey some of the flavor of the difficult bargaining that was taking place:

"We ran into quite a lot of deep water on oceans. We agreed not to discuss things we had already agreed on in prepcomm 3 and then ignored the agreement. There is an enormous lack of discipline in the delegations."

"The signals on cross-sectoral issues are bad. The financial discussions are bad. There are no conciliatory gestures."

"The discussion of toxic chemicals is worse than we expected. Delegates reopened issues that had already been agreed on in prepcomm 3."

"On the radioactive waste issue, the United States is isolated in not wanting any language at all. This is an issue that carries a great deal of emotional clout with the NGOs."

"Some say we will get agreement on forest principles. But not in our lifetime."

"The OECD [Organization for Economic Cooperation and Development] is unwilling to come out with a statement on finances. We are in a state of crisis already. We will not see progress on Agenda 21 unless there is progress on cross-sectoral issues. We need movement on finances."

"Biodiversity negotiations broke down last night. It is a crisis."

"It is a shame, but there is no agreement on high seas monitoring and enforcement or on protecting marine mammals."

"The Chinese are saying that negotiations on the atmosphere chapter [of Agenda 21] should not preempt the International Negotiating Committee negotiations [which was separately negotiating a climate change treaty]."

"There is a consensus on the environmental consequences of military activity. The only objection is from the United States, which does not want it included."

"The demographics discussion is not going well. The Holy See is cooperating directly with the chairman of the group, which is raising the suspicion of delegates. It is as if Exxon were in the chair of negotiations on energy."

"Everybody wants flexibility in the other person's positions."

"The U.S. delegation wanted to do away with the language about global consumption patterns. On consumption, the U.S. is pretty much alone."

"A minor revolt has broken out on institutions [which would oversee compliance with agreements reached at the summit]. Some are arguing for a new commission. Others want it to be done by ECOSOC [the Economic and Social Committee of the UN]. We may not be able to find common ground. We may have to carry it to Rio."

. . .

While the conflicting desires and demands of North and South were a leitmotif running through the entire process, there were also divisions within these major blocs, particularly in the Group of 77, although veteran diplomats said that, in general, the 77 displayed unusual solidarity during the lead-up to the summit. Shifting alliances and coalitions appeared within and between blocs. For example, led by Saudi Arabia, oil nations tended to present a united front against language in the energy and atmosphere chapter that would urge greater efficiency and the development of alternative fuels. Among Western industrialized countries there were a number of disagreements. One particularly bitter dispute persisted between Canada and several members of the European Union over "straddling stocks" of commercial fish that were important to both sides.

The breakup of the Soviet Union and the crumbling economies of Russia and its former satellites added another tense dimension to the negotiations. Previously, the Soviet Union had been a political ally of the developing countries on many issues and had been an important supplier of development assistance. Now the nations of the former Soviet Union were "economies in transition," themselves in desperate need of financial help as they tried to make the transition from centrally run state economies to a free-market system. As such, they were seen by the Third World countries not as allies but as rivals and competitors for the limited financial aid from the rich countries of the North. For the first time a number of newly independent countries, such as Armenia and Azerbaijan, which had been part of the Soviet Union, had their own official delegations and agendas at a UN conference.

Mikhail Kokeev, head of the Russian Federation's delegation, appealed for understanding. "We prefer," he said, "not to have any kind of competition with the developing countries. In the past they were helped substantially by my country. They will understand the serious nature of our problems in Russia now. We are beginning the deepest political and economic reform in our history. It is proving to be much more difficult than we thought."[6] But the dramatically altered politics of eastern Europe complicated the already tangled negotiations.

. . .

At one morning meeting halfway through the prepcomm, Strong said to Tommy Koh, "We need to turn fairly soon to a working program for Rio." To which Koh replied: "I don't know if we can finish this session successfully, so I don't even want to think about Rio."

. . .

Because of its size and complexity, the number of participants, and the volume of paperwork involved in the conference, even the practical details of running the meeting turned nightmarish. Translators were always in short supply, and there were constant complaints from delegations other than the English-speaking ones about not having documents they could read. Nor were there enough interpreters, particularly at the after-dinner meetings, which became more frequent and stretched farther into the night as the prepcomm raced to complete its work. At one evening session a delegate from every participating francophone nation got up and made a statement of complaint about the failure to have a French interpreter. Often there were not enough rooms available at the

UN to accommodate all of the formal and informal meetings taking place simultaneously. Many countries, particularly those of the Third World, had such small delegations that they could not send representatives to more than one or two meetings at a time. (No problem for countries like the United States, which had huge delegations backed up by virtually limitless government experts.) Delegates were often exhausted by meetings that lasted from early morning until late at night. Meetings had to be shuffled and rescheduled to accommodate delegations, room and interpreter availability, and social engagements of delegates. At one point, Tommy Koh, the chairman of the conference, stood by a bulletin board outside the plenary hall furiously scribbling meeting times and rooms with the stub of a pencil. He turned to a passerby and asked, "Phil, do you happen to have a Magic Marker?"

Some members of Strong's secretariat felt that the entrenched UN bureaucracy was one of the biggest hurdles that the Earth Summit preparations was forced to leap. One staff member recounted how, on his own initiative, he had rushed out with notes he had taken of the results of a negotiation, written them up, printed them out, and distributed them to three hundred delegates so that they could examine them before they decided whether to approve them. "One of the senior UN conference officials told me the next morning, 'You really shouldn't have done that. You are going to spoil them; the next time they will expect it.' And that's the attitude. That's the attitude. The purpose is not to deliver the information fast to the delegates; the purpose is to keep the system going. And unless you push them from the outside, it will never change."

For a while it seemed that for want of a nail the kingdom could be lost.

. . .

Hundreds of representatives of environmental, business, women's, youth, and indigenous peoples' groups took part in the final prepcomm, packing the seats reserved for them in the plenary chamber, swarming through the halls, serving on national delegations—the United States had more than twenty NGOs on its delegation at one point—presenting their own papers, speaking at plenary sessions, and holding caucuses. Arrangements had been made for "dialogues" between the NGOs and representatives of governments and UN agencies on such topics as oversight of transnational corporations. The secretariat, regional caucuses, and individual national delegations conducted special briefings for NGOs.

For the first time, representatives of environmental and grass-roots groups from the developing nations were present in substantial numbers. Some of them, like the O Le Siosiomaga Society of Western Samoa, with about one hundred members, were quite small.[7] Many of the nongovernmental participants were little prepared initially to take part in labryrinthine UN negotiation process, but all of them were militant in representing their positions.

To their chagrin, the environmentalists and grass-roots operatives from North and South often found themselves during the early prepcomms to be as sharply divided as their governments were. As Barbara Bramble of the National Wildlife Federation in the United States reported:

> We are split over styles, strategies, even language. For example, when we talk about "reallocation," we mean a good thing—like taking away fiscal incentives to cut down rain forests for cattle ranches. They use the word to mean a bad thing—taking current official development aid and relabeling it for the environment. We are asking for reforms of financial institutions. They are just asking for more money. They don't know that in the U.S., "wise use" is a code for the destruction of natural resources by big corporations. The split is a mile wide and 60 miles deep.

There were also sharp disagreements among northern NGOs, even among those from the same country. One of the bitterest arguments was between representatives of women's groups and population groups over language on family planning and on the role of women.

But by the middle of the fourth prepcomm, Bramble found, "some good things are happening. The doors between us and NGOs from the Group of 77 are breaking down. We are able to talk to them now. We are gaining the ability to understand each other's code words and what our real differences are. We can now talk to each other face to face and say what we mean, like we want reform in international banks but also in their own national institutions, too."[8]

Among other changes, the participation of NGOs in the Rio process was growing progressively more orderly. The dozens of organizations were increasingly speaking through caucuses, including the Third World Network and the Environmental Liaison Center, representing developing-country NGOs; the U.S. Citizens Network, representing grass-roots groups; and the Consortium for Action to Protect the Earth (CAPE), which included Friends of the Earth, the Sierra Club, the Environmental Defense Fund, the Natural Resources Defense Council, the National Wildlife Federation, and the National Audubon Society. Environmental groups in Europe joined in the European Environment Bureau. An important coordinating group was CONGO—the Congress of NGOs with consultative status at the UN. The International Facilitating Committee was a broad-based coalition organized just for the Earth Summit process. There were also a number of alliances on specific issues, such as the Climate Action Network.

Some of the American environmentalists, who felt that their participation in the preparatory process might have been little more than window dressing, came to believe that they were able to make some impact on the process, especially those who were included on the U.S. delegation. Elizabeth Barratt-Brown of the Natural Resources Defense Council, an observer on the U.S.

delegation, found that she was able to lobby delegates from other countries right on the floor of the plenary session. "My approach was pragmatic," she said. "It was to develop personal relationships with people. We convinced the head of the finance talks that there needed to be some reforms in the Bretton Woods institutions and he said, 'Get some language in to me.' We also got to some of the people on the U.S. delegation. A lot of the language on the Bretton Woods amendments has our fingerprints on it."

But Barratt-Brown also felt that the U.S. environmental organizations did not have as much impact on the course of the Earth Summit preparations as they might have had because "the major groups did not devote tremendous resources to this. Had we had an all-out effort, I think we could have gone to Congress early on and said, 'This is what the United States needs to do in Rio' and tried to get some legislation passed."[9] NGOs from other countries also perceived that their effectiveness was circumscribed. Geoffrey Greenville-Wood, president of the United Nations Association of Canada, noted that "on the most contentious issues between the North and the South, such as financial flows, institutions, trade and the environment, and transfer of technology—the political limitations of NGO influence become obvious. The instructions to the delegation come directly from the prime minister's cabinet."[10]

Ambassador Robert Ryan, deputy director of the U.S. delegation, who, at the time of the summit process was a veteran of thirty-four years of diplomatic service, said that the participation of "major groups" (using the term NGOs, he said, did not reflect the role of others such as business and women's groups) was "unique" in terms of the number and variety being consulted and serving on delegations. Their effectiveness, however, was "spotty," he felt.

> There were relatively few people, although the few were very good, who really knew how to make an input into negotiations effectively and how to serve effectively on a delegation. Quite a few people didn't really know how to focus on a text and what was happening in negotiations and, therefore, how to affect what was really happening. A lot of the inputs we got were at a very generalized level. I probably spent more time listening to people tell me that the U.S. should accept targets and timetables for reducing greenhouse gas emissions than any single thing. But we knew very well how the environmental community felt about that. They really weren't affecting things by telling us over and over again that particular thing.[11]

Negotiations on the global warming treaty, meanwhile, were being conducted by a UN-sponsored International Negotiating Committee on a separate track outside the preparatory committee process. They were not going particularly well. During his successful presidential campaign, George Bush had promised that he would answer the greenhouse effect with "the White House effect." Environmental activists were determined to "hold his feet to the fire" to keep his commitment, noted Rafe Pomerance, who, as president of Friends of the Earth and then as a senior associate of the World Resources Institute, had labored for over a decade to persuade the international community of the seriousness of the global warming threat.[12] But many business leaders, supported by the political right wing, opposed a treaty, fearing it would slow economic growth by mandating reduced use of coal and oil and increasing the cost of energy.

Only because of continuing pressure from the scientific community, environmentalists, and the media did the White House agree to engage in the negotiations. And the U.S. delegation made some valuable contributions to the negotiations, including proposals to limit emissions of all greenhouse gases, not just carbon dioxide, according to Pomerance. For much of the process, however, the United States acted as a drag on any movement toward a treaty. Pomerance, who was an observer and lobbyist for a strong treaty during most of the negotiations (and who later would be in charge of global warming issues as a deputy assistant secretary of state in the Clinton administration) said that John Sununu, the conservative White House chief of staff, was the controlling influence on the U.S. delegation.

Adamant opposition to the inclusion of binding targets and timetables for limiting the emission of carbon dioxide was the *sine qua non* of the U.S. negotiating posture. The industrialized countries of Europe were equally adamant about including such targets and timetables, at least in their public statements. But the United States had one huge bargaining chip: the decision as to whether President Bush would or would not attend the Earth Summit. U.S. officials made it repeatedly and abundantly clear that the president would not go to Rio unless there was a climate change treaty he could sign, and that meant no binding targets and timetables. Prepcomm 4 came and went with no commitment from Bush to go to Rio.

Negotiations on a treaty to preserve the planet's biological diversity were not proceeding all that smoothly either. Here again, the United States was a major stumbling block. A major source of friction was American resistance to demands by Third World countries that they be given concessional access to biotechnology developed from genetic materials originating within their borders. They argued that since plants and other genetic materials were the basis of much of the new drugs, foods, and other products of the new genetic technology, they should reap part of the benefits—benefits going entirely to the biotechnology firms of the industrialized countries.

But the American delegation contended that, if met, the demands of the Third World countries would diminish the value of intellectual property rights of the companies that had invested in the development of the genetically engineered products. Vice President Dan Quayle, who led the Bush administrations' efforts to roll back environmental regulations, wrote a memo attacking the draft biodiversity treaty, saying it would "hamper the U.S. biotechnology industry, greatly expand the reach of the Endangered Species Act and force the United States to enact a host of other regulations."[13] Buff Bohlen warned that "the chances are only 50-50 that the treaty will have satisfactory language on the issues fundamental to the U.S."[14]

Mostafa Tolba, the director of the United Nations Environment Program and a driving force behind the biodiversity treaty negotiations, insisted that the treaty had been drafted to protect intellectual property rights and the investments of industry. "But I understand," he added, "that this is an election year in the United States and it is not the right time to bring up anything that touches on industry or technology or property rights."[15]

. . .

By the beginning of the final week of the New York prepcomm, there was still no agreement on the most of Agenda 21 or on the Rio Declaration. While negotiations had produced results on many issues, governments from the developing countries were unwilling to give formal approval to any texts on protection of the environment without formal commitment by the industrialized countries to offer at least some new and additional financial resources and technology transfers to help them achieve environmentally sound development. The Group of 77 also sought the creation of a new "green fund" to disburse money to help them carry out the action programs of the agenda. The documents were littered with square brackets enclosing language to which one or more participating nation objected and thus required further negotiation.

While delegations from western Europe and Japan were giving at least rhetorical commitments to such financial support, the U.S. delegation offered not the slightest concession. The United States also resisted efforts by Third World countries to obtain language that pointed to the disproportionately heavy impact on the global environment made by the high consumption levels in the industrialized countries. President Bush had stated that "the American way of life is not negotiable," making it difficult in the extreme for the U.S. delegation to make any concessions on consumption. One senior member of the U.S. delegation said in a not-for-attribution conversation that "maybe this got off on the wrong foot with Maurice [Strong] stomping around the world raising the expectations of the developing countries. It raised the perception that we are being held ransom by the developing countries. They won't cooperate unless Uncle Sam pays."

Many participants in the process, including Strong and his deputy, Nitin Desai, credited the U.S. delegation with positive contributions to successful negotiations. "I have rarely seen the United States so active," Desai said. "On the fisheries issue, for example, the U.S. is trying to play the role of mediator. On institutions it has proposed a high level follow-up to Rio, unlike the United Kingdom, which has been much more conservative on this issue."

But increasingly, U.S. unwillingness to give ground on finances or consumption, its role in the climate and biodiversity treaty negotiations, and President Bush's coy refusal to commit himself to attending the summit were casting Washington in the role of villain in the Earth Summit drama. A front page *New York Times* article reported: "Rich and poor countries attending a five-week conference here say the United States is endangering prospects for the international environmental meeting planned for Rio de Janeiro next June. These countries say that by refusing to promise more environmental aid to the Third World or to set limits for emissions of gases that contribute to global warming, the United States threatens to sabotage the two most important agreements to be taken up at the Rio meeting."[16]

It should be added, however, that most members of the U.S. working delegation, led by Bohlen and Ryan, were highly respected by most conference participants and not held responsible for the major policy decisions that the United States brought to the bargaining tables.

As the final week of the prepcomm opened, Tommy Koh stepped briefly out of his role as dispassionate moderator and made a personal appeal to the delegates. Time was running out, he said. If a large number of unresolved issues were sent to Rio by the preparatory committee, they could not be resolved by heads of state. The members of committee must show the political will to reconcile their differences. "Let us not turn this prepcomm and the Summitt in Rio into a North-South confrontation. Instead, let us make the objective of Rio—generating economic growth in harmony with nature—a cause which unites us. Let us emphasize the positive rather than the negative . . . let us forge a new partnership between North and South, East and West, South and East for the 1990s and beyond."[17]

At an evening plenary session on the all-important issue of financial resources to implement Agenda 21, with negotiations still in deadlock, Ambassador Marker rose to make a conciliatory statement; he said that the developing countries were aware of the importance of the summit process and of the need to protect the global environment. But the process, he added, must be regarded as a "package" in which environment and development are linked. "The Group of 77 will continue to negotiate in the best possible faith in the spirit of compromise."

Then the delegate from China identified his government with Marker's statement, saying, "We believe that success in Rio is in the interest of all humankind. However, Agenda 21 cannot be implemented without adequate financial resources. A housewife, no matter how competent, is unable to produce a rice pudding without rice. We want a signal from the industrial countries conducive to confidence building. It doesn't have to be a specific number."

Shortly thereafter, Buff Bohlen of the United States took the floor. The United States recognized, he said, that financial resources and mechanisms would be needed if there was a "general agreement" on Agenda 21 and on issues such as forests, marine species, national capacity building, and reducing pollution. Then he added: "Of course industrial countries must generate new and additional resources from the public and private sector and reallocation of existing resources. There is no question that developing countries and countries in transition must have new resources. I would like to make it absolutely clear that the United States is committed to working with other industrial countries to mobilize new and additional resources for a new partnership."

Bohlen's statement had an immediate, electric effect on the meeting. The United States had, at last, committed itself to the principle of new and additional resources to help the poorer nations achieve sustainable development! An excited buzz of voices broke out on the floor, and a small crowd of delegates from other nations rushed up to Bohlen to congratulate him. Speakers from the European Union, Japan, and Sweden rose to associate themselves with Bohlen's remarks. Ambassador John Bell of Canada, who was chairing the negotiations on financial resources said, "I have never been so encouraged."

As it turned out, U.S. support for new and additional financial resources was wholly rhetorical. The Bush administration was not prepared to allocate a single additional penny to development aid. If there was to be any new money from America, it was going to come from the private sector not the government. "We were in an impossible position because we had nothing to put on the table," a senior member of the U.S. delegation later explained. "More aid was a non-starter given the mood of the country and the recession. And then [ultraconservative ideologue Pat] Buchanan was pushing from the Republican right. This exercise could not have come at a worse time. Clearly you wouldn't pick a presidential election year."

So amorphous a commitment on financial resources could not and did not break the negotiating logjam on finances. The prepcomm ended without agreement, and the question of new and additional resources was forwarded to Rio for resolution at the top political level.

But Bohlen's demarche did seem to give the prepcomm a strong jolt of energy. Negotiations moved forward swiftly on a number of fronts. After marathon and sometimes emotional sessions, prodded by the heavy gavel of chairman Koh, the negotiators reached consensus on the draft text of the "Rio Declaration on Environment and Development." It would be the only text to go to the summit without brackets—there were no objections to any of the language. (Israel had meant to object to a reference to "people under oppression, domination and occupation," which was taken as meaning the Palestianians, but was gaveled down by Koh.)

The declaration did not turn out to be the short, ringing affirmation of support for the planet, the Earth Charter, sought by conference organizers and many of the northern countries. As noted in the *Earth Summit Bulletin*, a painstakingly prepared summary of the daily proceedings at the prepcomm, and later in Rio, it was "an attempt to balance the key concerns of both Northern and Southern countries." It contained twenty-seven "norms for state and interstate behavior, many of which have never been universally accepted before."[18] Among other things, the South obtained a commitment to the eradication of poverty as a prerequisite of sustainable development, a recognition that humans are "at the center of concern for sustainable development," support for an "open international economic system," and an affirmation of their "sovereign right to exploit their own resources." The developing countries also were able to insert language, resisted by the United States and other northern countries, asserting that "states have common but differentiated responsibilities" for achieving sustainable development and that the developed countries had a special responsibility "in view of the pressures their societies place on the global environment and of the technologies and financial resources they command."[19]

For the northern countries there was language on transboundary pollution; the development of environmental liability law; application of the "precautionary principle," which requires that environment threats be addressed even in the absence of conclusive evidence; and the "polluter pays principle." The text also acknowledged the role of women, youth, and "indigenous people" in achieving sustainable development.[20]

General consensus also was achieved on a draft of Agenda 21. The draft, however, remained thick with square brackets representing failures to agree on many of the specifics, most particularly on financial arrangements and North-to-South technology transfers. Because consensus was required to forward the document to the summit, many of the provisions were relatively weak reflections of the lowest common denominator among the participating governments. On some agenda items, such as the "Authoritative Statement of Forest Principles," an issue that raised the fears of such countries as Malaysia and India about threats to their sovereignty, the language was so weak as to be virtually meaningless. There were gaping holes in agenda items dealing with help for the impoverished, despite Strong's repeated insistence that the eradication of poverty must be a central goal. The draft document also did not fully come to grips with population, fisheries, disposal of radioactive wastes at sea, liability for transboundary pollution, the use of energy and other consumption issues, and dozens of other concerns.

Strong's deputy, Nitin Desai, a highly regarded economist from India, felt that the failures in the negotiations to a large degree reflected the political conservatism that was dominating the United States, Japan, and much of western Europe. Appeals to morality, compassion, and responsibility no longer moved the political power structure of those nations, he said. "We forget we are not talking to the liberal establishment in the West anymore. We have to get across why the rich should care about the poor, even poor people sitting in Brazil or Nigeria. We have to demonstrate that the consequences of poverty are mass migration, deforestation, disease, drugs and political insecurity. The only way to move the conservative establishment is to convince them that poverty is a threat to them."[21]

Nevertheless, Desai felt, the achievements of the preparatory committee represented real progress:

> Part of what we are trying to achieve here is the process; we are trying to get countries to act internationally. The goals of sustainable development involve major compromises of sovereignty. There is no commitment yet to diluting sovereignty. Before that happens we must have a credible system of international governance. But our job is not to do just what will happen in our lifetime. This process is an important step toward global governance, in which governments will have confidence and will surrender sovereignty.

Japan's Noburo Takeshita had yielded to Maurice Strong's blandishments and in mid-April acted as host to an Eminent Persons' Meeting on Financing Global Environment and Development in Tokyo. Former U.S. president Jimmy Carter was among the eminent persons, as were former presidents of Costa Rica, Mexico, Singapore, Nigeria, and Colombia. Also present were personages from the world of diplomacy, finance, business, and industry and, of course, Maurice Strong. While the meeting was not expected to produce any specific commitments of money and did not do so, it did put the prestigious gathering on record as supporting the mobilization of financial resources for sustainable development. It also, Strong hoped, placed Japan in the van of the financing struggle and perhaps would induce the Japanese government to jump-start the process with a substantial commitment.

. . .

A month after the end of prepcomm 4, the International Negotiating Committee on climate change, also meeting at UN headquarters, reached agreement on a compromise treaty to address the global warming threat by reducing emissions of greenhouse gases. At the insistance of the United States, which alone opposed them, the draft treaty set no specifically linked targets and timetables for reducing emissions. The agreement was a triumph for U.S. negotiating strategy, which had held out the threat that President Bush would boycott the Earth Summit if its terms were not met. With the compromise it was assured that the president would go to Rio.

There was some question as to whether Bush's threat to stay at home was ever credible, particularly in a year in which he was running for reelection. Most world leaders, including U.S. allies, had already announced they would go to Rio, whether or not Bush came. As an official on his campaign staff noted, "You don't want to be the only guy on the planet who doesn't show up." Environmentalists conducted an extensive campaign to persuade the president to attend the summit, including a flood of letters to the White House. As the *Wall Street Journal* noted, Mr. Bush, who had promised four years earlier to be the "environmental president" was "in danger of being tagged in Rio as No. 1 enemy of the Earth."[22]

Many environmentalists, as well as European environmental officials, expressed disappointment over the draft treaty. "It will be possible for anyone to wiggle out from any legal commitment," complained Jorgen Henningsen, environmental director of the European Commission. But as the *Washington Post* pointed out, "by embracing broad goals and initiating a process for regular updating of emission standards, developed nations took a first step in what could go well beyond previous efforts to protect the global commons."[23]

Two weeks after agreement was reached on the global warming treaty, negotiators in Nairobi approved the draft of a convention to protect the planet's biological diversity. The way was cleared for the treaty when agreement was reached on language that would assure the developed countries continued access to genetic materials in the developing countries, and those countries would be assured "fair and equitable" access to biotechnologies developed in the industrialized nations. The United States had reservations about whether the language of the treaty would protect the intellectual property rights of companies that had developed the biotechnologies but gave no indication in Nairobi that it would reject the treaty.

. . .

By mid-May, therefore, just two weeks before the scheduled opening of the summit meeting, against the predictions of the pessimists, major treaties on climate change and biodiversity were ready to be signed in Rio. The final prepcomm, which one participant described as "a diplomatic marathon that belongs in the *Guinness Book of Records*,"[24] had approved of the two key documents for the summit: Agenda 21 and the Rio Declaration on Environment and Development. At least one hundred heads of state or government, including the president of the United States, had announced their intention to attend the conference. While the preparatory process had underscored many of the weaknesses of the UN process—infuriating bureaucracies, endless rhetorical posturing, inability to provide leadership unless a powerful government took the lead—it nonetheless had proved adequate to the task of achieving consensus among the nations on what was arguably the most urgent and intricate set of issues facing the global community. The table had been set for Rio.

There was, however, little euphoria among members of national delegations, the conference secretariat, and environmentalists as they packed their bags for Brazil. Much unfinished business remained to be completed at the summit, including putting flesh on the somewhat bare bones of the action agenda, obtaining commitments for the financial and technological means to carry out the agenda, signing the treaties, creating an institutional follow-through, and mobilizing the political will to assure the decisions of the summit would be translated into policy from the global to the local level. It was by no means certain that the spirit of cooperation that had belatedly rescued the preparatory process would carry over to the summit, to which presidents and prime ministers would bring their own political agendas.

It was not even clear that the Brazilian government, entering a crisis created by charges of corruption against its president, would be ready to receive the thousands of officials, nongovernmental activists, and journalists about to descend on Rio.

So, with apprehension as well as with hope, the eyes of the world turned toward Rio de Janeiro at the beginning of June 1992.

Chapter 11

At the Summit

O Zion, that bringest good tidings,
get thee up into the high mountain;
O Jerusalem, that bringest good tidings,
lift up thy voice with strength;
lift it up, be not afraid;

—Isa. 40:9

In terms of sheer magnitude, there had never been anything quite like the United Nations Conference on Environment and Development. With 115 heads of state and government attending, it was, as one participant quipped, "the mother of all summits." In all, 178 nations sent some seven thousand delegates to Rio. The meeting was covered by nearly nine thousand journalists, making it, according to the UN Department of Public Information, the most heavily reported single event in history. Over 1,400 nongovernmental organizations (NGOs) were represented at the conference,[1] and virtually all of them participated in the Global Forum, the "parallel" NGO forum assembled in Flamengo Park on the ocean front in downtown Rio. An estimated twenty thousand environmentalists and representatives of women's, youth, indigenous peoples, business, labor, religious, and other independent groups attended the forum, and many of them traveled to the official summit to speak, lobby, and brief reporters.[2]

The official conference was held at Rio Centro, a sprawling conference center that was not "centro" at all but located some ten miles away from downtown in the modern and relatively affluent suburb of Barra di Tijuca. Surrounded by tanks and machine-gun-mounted jeeps and guarded by grim-faced young Brazilian marines clutching automatic weapons, the conference center was a city unto itself, with auditoriums, blocks of offices, banks, shops, bars, restaurants, medical facilities, and a sophisticated telecommunications complex. During the meeting as many as twenty-thousand people at a time—

delegates, NGO lobbyists, reporters, conference staff, maintenance and service workers—milled around the cavernous buildings and along the open-air corridors.

To some, the size and complexity of the summit, the colorful and even festive surroundings at Rio Centro and the Global Forum, the stream of Very Important People with their retinues, the hordes of journalists and unofficial participants, the picturesque national costumes, suggested nothing so much as a three-ring circus, a kind of summertime carnival in Rio. But on the eve of the conference (its opening had been put back two days to June 3 and the summit itself to June 14 to avoid conflict with a Moslem holy day), arriving delegates quite clearly were gripped by a sense of serious purpose.

Celso Lafer, foreign minister of Brazil and vice president of the meeting, stated the day before its opening that "this is not one more conference. We have a clear mandate from the international community to begin to create a new order for the well-being of mankind."[3]

The conference "sherpas," while accomplishing a great deal, had not entirely cleared the path to the summit. Much remained to be resolved in Rio, and the outcome of the meeting was by no means certain. If thorny issues such as financial measures, technology transfers, forest principles, and follow-up institutional measures, as well as a host of less crucial but still divisive issues, were not resolved, the summit could turn out to be, as one observer put it, no more than "a global photo-opportunity for a group of leaders who don't deserve it."[4] Delegates canvassed on the eve of the summit were split "between the pessimists, who said they were girding themselves for the possibility of a diplomatic debacle of unparalleled proportions, and the optimists, equally convinced that whatever disappointments may emerge in the next two weeks, the Summit will have been a vital first step toward rescuing the environment."[5]

The opening ceremony for the conference was held on a vividly green lawn outside Rio Centro as a brilliant sun shone on the dignitaries and a scarlet-and-white-clad military honor guard stood at rigid attention. But the opening remarks by Boutros Boutros-Ghali, the new secretary-general of the UN, a tall, saturnine figure in the midst of this refulgent scene, were somber: "This is a time of a finite world. It is a time we are under house arrest. Henceforward, nature is in the hands of man. Man has triumphed over nature. Every new triumph will be progress against ourselves. . . . We are condemned here to move closer together to a more virtuous planet. . . . The new collective security is a planetary development. It is now less of a military than an economic and ceological issue."

. . .

Maurice Strong, in his opening charge to the delegates, called urgently for stabilizing global population and ending world poverty, two issues that had received relatively short shrift in the preparatory meetings—and would not fare much better at the conference. But, he warned,

> We are all in this together. No place can remain an island of affluence in a sea of poverty. No part of the world can live in the indulgence of unrestrained consumption. . . . North-South, rich-poor, we must have a new global partnership for a more secure and hospitable planet. . . . The damage to the planet has been largely inadvertent. But we now know what we are doing. We have lost our innocence. If we lapse into business as usual we will have lost our historic opportunity to bequeath a better world to the future.

President Bush, however, already had put something of a damper on the conference when the White House announced a couple of days before the opening that the United States would not sign the treaty to preserve the planet's biological diversity. Administration officials said that the treaty would weaken the patent rights of American biotechnology firms and would require an expansion of the Endangered Species Act, a law disliked by American conservatives, who complained that it infringed on private property rights.[6] Annoyance at the United States intensified when a staff member of Vice President Dan Quayle's antiregulatory office leaked a memo from U.S. Environmental Protection Agency administrator William K. Reilly, which suggested that the White House would be able to accept the treaty with a few minor changes. The White House rejected the suggestion, and the memo had obviously been leaked to embarrass Reilly, a conservationist continually at odds with the right-wing conservatives in the administration. U.S. officials compounded the annoyance by trying to pressure some of its allies, including Canada, the United Kingdom, and Japan, to join in boycotting the treaty. As it turned out, the United States alone among the major industrial nations refused to sign. Before Rio ended, more than 150 nations had affixed signatures to the treaty, which requires national plans for preserving species and their habitats, funds to help the developing countries safeguard their biological resources, assured commercial access to resources, and fairly shared revenue from the commercial marketing of products developed from those resources.

Even before the conference opened, there was strong criticism of the Bush administration's insistence on a climate treaty with no numerical goals and for holding the treaty hostage to the president's decision on whether or not to go to Rio. The European Community's environment commissioner, Carlo Ripa di Meana, announced that he would not attend the Earth Summit because of his disappointment with the weakness of the treaty.[7] The poorer countries were incensed by Washington's continuing refusal to commit new and additional resources to help them on the path to sustainable development.

To the media at least, the Bush administration had quickly managed to transform a complicated international negotiation into a simple morality play.

For several days, the story of the summit in newspapers and on television programs around the world was the good guys trying to save the world and the bad guys trying to stop them, with the United States cast in the role of the bad guys. A front-page headline in *Jornal do Brasil*, a leading Brazilian paper, said, "US Singled Out as Eco Bad Guy."[8] The *Earth Summit Times*, the official newspaper of record for the conference, spoke of "the American slide into isolation at the Earth Summit."[9] Sir Shridath Ramphal, a Guyanan and former secretary-general of the Commonwealth of Nations, said in a briefing that he regarded the U.S performance in Rio "with sadness" because in the past America had led the way to internationalism and now was moving in the opposite direction. Nevertheless, he added, "one player, no matter how powerful, cannot keep the game from being played."[10]

Sir Shridath was correct. As they waited for their national leaders to arrive in Rio, negotiators moved forward on a broad front to eliminate remaining differences in the text of Agenda 21. As one member of the secretariat staff observed, "No nation wants to be responsible for a failure in Rio." The question of a follow-up institution, which had seemed almost unobtainable in New York, was resolved with an agreement to ask the General Assembly to create a fifty-two-member Commission on Sustainable Development, which would monitor compliance with the Rio agreements and would report to the UN secretary general through the Economic and Social Council. Disputes over the chapter on the transfer of environmentally sound technology were ended with hedged language that spoke of "promoting" rather than ensuring access to technology for the developing countries and "measures to prevent the abuse of intellectual property rights" of corporations in the industrialized countries. After the United States reconsidered and decided not to object, the delegates agreed to recommend that a treaty to prevent the spread of deserts be negotiated in the near future. The U.S. delegation also decided not to demand the reopening of the Rio Declaration to remove what it considered to be objectionable phrases, such as "the right to development," and instead opted to issue its own separate "Interpretative Statements for the Record." While many considered the U.S. statement somewhat less than edifying, it did make it possible for the summit to move forward. A reopened declaration might have turned into an unclosable Pandora's box.

At his morning staff meet as the second and final week of the conference opened, Strong said that "things are off to a promising start. Negotiations have gone a good way toward completing the issues left open." His deputy, Nitin Desai, added: "Agenda 21 is practically complete. We are down to key political issues."

The most difficult unresolved issue remained the crucial question of finances—money to assist the developing countries achieve the goals of sustainable development. The Group of 77 still insisted that its own draft be the basis for negotiation of the financial resources chapter of Agenda 21, which included such measures as a new global fund and which virtually all the industrialized countries found largely unacceptable. Seeking to break the impasse, Tommy Koh, chairman of the main committee of the conference, asked Rubens Ricupero, the Brazilian ambassador to the United States to coordinate the negotiations. He offered his own draft chapter, a combination of goals sought by North and South. At the heart of his proposal was a recommitment of the industrialized countries to a long-standing pledge by most of them to devote 0.7 percent of their gross national product to development assistance for the poorer nations. The new plan helped break the gridlock on negotiations, but agreement was not reached until the eleventh hour of the conference.[11]

Agreement on meaningful forest principles also proved to be elusive, even though they would not be legally binding and despite a strong push by the United States for its own plan to slow the destruction of tropical forests. In often "acrimonious"[12] debate, many of the developing countries, led by Malaysia, objected to efforts to place limits on how they use their forests as an infringement of their sovereignty. "If the developed world thinks the forest is an [international] heritage . . . they should pay for its conservation," Malaysian ambassador Ting Weng Lian told an interviewer from his country.[13] Because the industrialized countries had failed to set targets for reducing their emissions of carbon dioxide, the Third World nations resisted having the focus placed on their forests as sinks for capturing the greenhouse gas. The rich countries, meanwhile, showed little enthusiasm for paying for the preservation of tropical forests in the South. A U.S. pledge of $150 million for forest protection was dismissed as a sop.[14] What emerged was set of principles that environmentalists and even forest industry lobbyists regarded as so toothless as to be meaningless. John Heisenbuttel of the American Forest Council, an industry group, noted that the principles did not even contain a definition of "deforestation."[15] And while some had hoped that the principles would be the precursor of an eventual binding treaty to protect the world's rapidly diminishing forests, the agreement in Rio studiously avoided any recommendation of further action.

Another series of pitfalls on the way to full agreement on Agenda 21 was created by Saudi Arabia, usually joined by Iran and Kuwait, which sought to eliminate all references to fuel conservation and alternative sources of energy in chapters on energy, protection of the atmosphere, and technology. The Saudis, all too obviously trying to block any agreements that would enable the world to lessen its dependence on oil, conducted what, in effect, was a filibuster against a consensus on those chapters. Discussing the Saudi intransigence at one morning staff meeting, Strong complained that the requirement that agreements be reached by consensus was "the real veto power in the United Nations" and said that on this issue he might have to call for a vote. Nitin Desai interjected that "the fact we are bending over backwards to accommodate all points of view will help us in the end." But Strong said, "I am willing to risk my credibility on this issue. We just cannot have wishy-washy language on energy and energy efficiency at this conference." Strong, who had studiously avoided public criticism of any participating nation, including the United States, indicated that he might take the rostrum to denounce this stance.

By and large, however, the table was well set for the heads of state and government as they arrived for the summit proper on June 12. "We are very close to home," Strong told reporters on the eve of the meeting.[16]

. . .

The summit itself was an epic oratorical marathon. More than one hundred national leaders, each limited to seven minutes, took the rostrum of the plenary hall, many of them quite obviously speaking to audiences in their own countries for their own domestic political purposes. Within this rhetorical hothouse, however, blossomed a coherent, unexpected theme of global solidarity—a kind of "comrades in the foxhole" sensibility arising from the recognition of shared danger in the face of a rapidly deteriorating human habitat. Virtually every speaker, North or South, from every continent, acknowledged that his or her country was in the same leaking lifeboat with every other country and only cooperation would save them. In the preparatory process, the industrialized nations had been concerned primarily with the environment, the poorer countries with economic development. It was almost as if they were preparing for two different conferences. At the summit the two goals merged, at least in the words of the national leaders.

The tone was set by the very first head of government to speak, Prime Minister P. V. Narsimha Rao of India, who urged the nations of the world to "join hands" to reverse the "degradation" of the planet. It is clear, he said, "We cannot have conservation of the environment without the promise of development, even as we cannot have sustained development without the preservation of the environment. The recognition of this symbiosis is the only enduring basis on which this conference can attain its purpose."

A number of the speakers said that the summit meeting could pave the way to a new, post–Cold War relationship among the nations of the earth. Li Peng, prime minister of the People's Republic of China, noted that although the world was in a period of transition, "two major issues—peace and development—have not been resolved. Likewise, to address the global issues of environmental protection and economic development, cooperation will be required. The Rio conference and its conventions should lay a good foundation."[17]

Nevertheless, it was abundantly clear that the traditional *realpolitik* of suspicion, narrow national interest, and ideological belligerence was also present in Rio. The old world, the world of the past half-century, was captured in a brief tableau as the national leaders prepared to sit down to lunch together. President Bush, his lanky frame easily spotted as heads of state and government milled around the block-long table before sitting down, stood a few feet from Fidel Castro, also readily identifiable by his neat military uniform and his bushy, now silver beard. The two old antagonists, the former CIA director and the one-time revolutionary, were very obviously doing their best to avoid face-to-face confrontation or even eye contact.

In their speeches to the summit, both Bush and Castro displayed a dismaying failure to grasp the new possibilities open to the human community and represented by the Rio gathering. Castro, a world-class orator, spoke for less than five minutes, probably the shortest speech of his life. His rhetoric was that of the Cold War, blaming the destruction of the environment on "colonial metropolises and of imperialist policies that also engendered the poverty and backwardness which are today the scourge of the overwhelming majority of mankind." His remarks drew thunderous applause, but it might have been for their brevity rather than their content.

President Bush's speech, rather than offering any statesmanlike vision of the future of the community of nations and the American role in that future, was largely a defensive justification of the U.S. nay-saying approach to biological diversity, global warming, and helping poor countries finance sustainable development. The United States, he said, need apologize to no one because "its record of environmental protection was second to none." True or not, the statement was a *non sequitur* in the context of a conference setting a future course for international relations. And he neglected to mention that most of the environmental progress in the United States was initiated before he and his predecessor, Ronald Reagan, were in power. He also defended his decision not to sign the biodiversity treaty, saying it would lose American jobs. Implicity acknowledging that all of the other industrialized countries had agreed to the treaty, Bush said, "It's not easy to stand alone, but sometimes leadership requires we stand alone." Aside from its breathtaking banality, his remark also managed to insult virtually every other leader on earth by implying that he was the only one wise enough to reject a bad treaty.[18]

The *Earth Summit Times* reported: "The eagerly awaited speech by President George Bush at history's largest diplomatic gathering left many delegates and nongovernmental groups crestfallen and others simply disgusted."[19] Many participants commented that the United States, now indisputably the most powerful and influential nation on earth, was letting a historic opportunity to build the foundations of a new, cooperative international regime slip away. Senator Al Gore, who was leading a Congressional delegation to the summit—and who would several weeks later be tapped as the Democratic Party's vice presidential candidate—said that Bush's performance was "doubly tragic because it is so obvious that change is needed. Every nation in the world is looking to the United States for leadership."[20] Jean Musitelli, spokesman for President François Mitterrand of France said that "the Americans talk about the new world order but they'd be more credible if they heard the world's message on the environment, which will be one of the essential themes in this new world order."[21] European Community environment commissioner Ripa di Meana complained that "President Bush's declaration struck at the heart of the Rio Conference by putting out a series of negative signals . . . the largest superpower refuses to assume its environmental responsibilities."[22] And Emil Salim, Indonesia's minister for environment and population, a seasoned diplomat who had represented his country in Stockholm, lamented: "Of the major powers, only the United States has not risen to this global opportunity. The U.S. could really uplift the quality of this country and lead us into the twenty-first century. Tragically, the U.S. was boxed in by its elections into a short-term view and missed the opportunity."[23]

As he was about to depart from Rio, a Brazilian journalist asked President Bush about criticism of the United States and his performance. He answered: "I'm President of the U.S. not President of the world and I will do what I think is best for the U.S."[24]

At one point, the national leaders—not including President Bush, who had already left—met for a closed private session. Maurice Strong, conscious of the rare opportunity to address the assembled chiefs of state and government, told them that they "would be known to history more by what they did or failed to do in Rio than for all the other issues that might be preoccupying them at the moment . . . that they were acting as trustees for the future of life on earth, which was an awesome responsibility."[25]

But no other nation or bloc of nations stepped forward in Rio to fill the leadership vacuum left by the United States. German chancellor Helmut Kohl roamed the planary hall like a big jovial bear, content that the little his country—and indeed, the European Community as a whole—had done to make the summit a success was substantially more than the United States had contributed. But Germany was too preoccupied with the onerous task of unification to take a global leadership role in Rio. Strong had been counting heavily on Japan to help move the international community toward the goals of the summit, particularly the financial goals. While he reiterated that Rio was not to be a "pledging conference," he believed he had successfully prodded Japan to prime the pump with a commitment of substantial new and additional financial assistance to the developing countries. It was expected that Prime Minister Kiichi Miyazawa would announce a substantial increase in his country's development assistance when he addressed the plenary.

When the day of the summit arrived, however, Miyazawa did not. He was tied up by a domestic political crisis over whether the country's constitution would permit it to send troops abroad for peacekeeping missions and was forced to cancel his trip at the last minute. He did, however, send a videotape of his planned speech, in which he said that Japan would immediately increase its development aid by some $1.4 billion a year and seek to permanently expand the percentage of its GNP it gives to developing countries. The money was not as much as Strong had hoped for, but he felt it represented an important step forward. He badly wanted to show the videotape at the plenary session or immediately following it, believing that the pledge of money would inspire other countries and reinvigorate the conference.

In a tense and angry confrontation with Strong, however, Secretary-General Boutros Boutros-Ghali vetoed the proposal to show the tape to the plenary session and even forbade it to be shown informally after the session concluded. It would, Boutros-Ghali contended, set a bad precedent that would encourage national leaders to send videotapes instead of attending future UN summits in person. The Miyzawa video was not shown. The Japanese leadership role in international affairs did not materialize. The other industrialized democracies were not moved to increase the trickle of development aid from their treasuries.

Strong was furious. He told Boutros-Ghali that he was deeply disappointed with the secretary-general's failure to support him. Later, Strong and Boutros-Ghali repaired their relationship. But to Strong, "the entire Rio experience made it clear to me that some of the most effective work in following up and implementing the results of the conference would have take place outside the formal U.N. system."[26]

. . .

As the summit moved toward its conclusion, Strong was a blur of activity, lobbying delegates for the agenda prepared by the secretariat, serving as chief publicist for the conference with one media interview after another, and acting as father confessor and a shoulder to weep on for diplomats and staff members. A member of the U.S. delegation said after the conference, "I always suspected that Maurice had found some new biotechnology to clone himself. I don't know how he managed to be in as many places as he was." During the three days of the summit he received an unending stream of chiefs of government and state, each bearing words of congratulation as well as suggestions, demands, complaints, and tales of woe. Perhaps none was more woeful that of the sad-eyed president of the recently independent nation of Armenia, Levon Ter-Petrosian, who told Strong that even as they spoke an Armenian city was under armed attack by Azerbaijani forces. But he had decided to come to Rio, he said, because environment and development problems were so urgent in his country. "Armenia, like all other states of the ex-Soviet Union, is in a state of ecological catastrophe. For a long time we will be unable to invest in our ecology because even the elementary needs of our people, including food and housing are not being met. We will not be able to take care of our ecology without substantial help from the West." Strong could only reply that he understood and sympathized and would do his best to get as much help as possible.

Finally, the last speech was delivered and the last brackets were removed from disputed clauses of Agenda 21. The tortuous negotiations on financial resources and mechanisms were concluded just before 7:00 p.m. on the last official day of the conference. Under the compromise (to say the least) wording, the developed nations reaffirm their commitment to the long-standing UN goal of allocating 0.7 percent of their GNP for official development assistance. But the wording did not actually commit the industrialized nations to increased levels of aid. There was no mandatory target date; the year 2000 was only a suggested goal. The United States took the position that it had never committed itself to the 0.7 percent goal and therefore was under no obligation to recommit. There would be no "earth increment" to the International Development Agency, the soft-loan window of the World Bank. Nor would there be a new, independent "green fund" to finance environmentally sound development. The fund had been sought by the developing countries, which disliked dealing with the World Bank. Instead, negotiators agreed to restructure the Global Environment Facility, a joint project of the World Bank, UN Environment Program, and UN Development Program as a channel for sustainable development funds and to make it more democratic by giving recipient nations a greater voice in its management.

Actual pledges of money for sustainable development made at the conference were estimated at between $3 billion and $6 billion, but the wording of the pledges made it difficult to determine how much, if any, was "new and additional." While Strong had said repeatedly that the summit was not a pledging conference, he was disappointed that the industrialized countries had not made more expansive demonstrations of economic cooperation to back their concerns about the environment.

The Saudis continued to try to block any reference to energy conservation and alternative fuels until the end of the conference, when Tommy Koh gaveled them down to the loud applause of the plenary session.

Otherwise, Agenda 21 was a product of the consensus of all 178 governments participating in the Rio conference, and thus many of its provisions represented what a number of participants and observers called "the lowest common denominator" to which national governments were willing to commit themselves. Although it is a nonbinding agreement, the calls to action in many chapters were so faint as to approach inaudibility. Because of President Bush's dictum that the "American way of life is not negotiable," for example, reference to restraining consumption could be interpreted as imposing virtually no obligations to reduce resource use. Lobbying by the Vatican and its allies in Latin American nations made sure there was not a tooth in the chapter on population issues. Language on transnational corporations, on the military causes of environmental degradation, and on ending poverty was diluted to weak gruel.

Still, Agenda 21 (400 to 800 pages long, depending on the kind of print used) was complete, the Rio Declaration was accepted, and the climate and biodiversity treaties were signed by most of the attending nations.

. . .

The Global Forum, the alternative summit, also came to a productive conclusion. An innovative experiment in extragovernmental diplomacy, the forum, like the official summit was the venue of endless speeches, lobbying, technical meetings, strategy sessions, document drafting, North-South confrontation, and eventual compromise and agreement. The independent sector in Rio produced some thirty alternative "treaties," including treaties for preserving biological diversity dealing with global warming, that were far more pointedly worded than the official treaties. They also agreed on an "Earth Charter," the kind of short inspirational document on the need to protect the planet and its human and nonhuman inhabitants that Strong had hoped for but had not achieved in the Rio Declaration. The NGO documents had no legal standing, of course, but they did establish standards by which to judge the output of the official summit and, more important, the follow-up actions of national governments. Maurice Strong gave forceful support to the forum and its organizers, braving Rio traffic on several occasions to make appearances at Flamengo Park. In all, the forum attracted an estimated one hundred thousand visitors while it was open.[27]

At the end, governmental delegations completely dominated the summit. Representatives of the independent sector were pushed, gently it is true, to the periphery of negotiations. But there can be little doubt that the nongovernmental activists had exerted at least some influence over the deliberations and decisions of the official delegates to the conference. The national leaders knew that the environmentalists and other nongovernmental participants would return home to report to their constituents after the conference was over. The NGOs' views had to be heeded for domestic political reasons if for no others. When, for example, some of the industrial nations, expressing distaste at the thought of creating a new bureaucracy within the UN, were considering opposing the Commission on Sustainable Development, they were at least in part dissuaded in heated meetings with the nongovernmental groups.

It was also largely because of private-sector demands that the interests of some members of the human family that are normally brushed aside at such diplomatic gatherings—women, youth, indigenous peoples—were included on the summit agenda. Led by former U.S. Congresswoman Bella Abzug, sporting her usual array of brightly colored hats, and Wangari Maathai of Kenya, feminist activists from around the world were omnipresent at the conference, bludgeoning women's rights issues onto Agenda 21 and forcing recognition that women are the primary caretakers of life on Earth and the foundation of much of its economic life as well. The final agenda called for the removal of legal, educational, cultural, and "attitudinal" obstacles faced by women and for their greater representation in decision making at all levels.

Representatives of indigenous cultures—tribespeople and forest, tundra, desert, and island dwellers—set up their own village in Rio and held their own "world conference" just before the summit to dramatize their role as caretakers of nature and their desire to maintain their traditional way of life, which was threatened on all parts of the planet by the pressures of the global economy. While some countries, including Canada, were wary of diluting their own sovereignty by granting special status to indigenous peoples in their midst, the rights of those people to be involved in decisions that affect their lives was recognized in Agenda 21.

For the nongovernmental activists, perhaps the most important accomplishment in Rio was the forging of something close to a sense of common identity. Environmentalists from all over the world, with very different conditions in their home countries and often with conflicting ideologies, had agreed on broad agendas for future action and created global networks for continuing communication and future action. Together they had come to believe that they constituted a new force in the world that could influence the course of events even without the participation of governments.

This sense of identity and potential power was summed up by Rosiska Darcy de Oliveira of the Brazilian Women's Coalition: "Now we are beyond government. There are new actors and NGOs are among them. There is a new kind of diplomacy, another kind of representative—they are the NGOs."

Many of the private-sector activists were disappointed with the outcome of the official summit, however. After the final gavel, Martin Khor of Malaysia, a leader of the Third World Network, wrote: "Most non-governmental participants and even, one suspects, many government delegates, feel the summit section of the conference has been all show and photo opportunities and little of substance. . . . Rio revealed that the Northern governments are not willing to yield the immense powers they have over world resources and the international economy."[28]

. . .

When asked to assess the results of the Earth Summit, Maurice Strong likes to tell of a conversation between former secretary of state Henry Kissinger and Mao Zedong when Kissinger traveled to Peking to arrange a meeting between the Chinese leader and President Nixon. After listening to Mao talk about the Communist revolution in China, Kissinger asked him what he thought had been the results of the French Revolution. Mao replied: "It's too early to tell."[29]

After the conference ended, Strong felt that only time would tell whether the Earth Summit would produce any historical changes in the way the human community ordered its affairs. The conference itself, he said, was only a beginning. But he was convinced that it had been a *political* success in that it attracted global political "engagement" by actors ranging from heads of state to local grass-roots activists. It was also a political success, he contended, because it finally forged the link between the environment and economics, between poverty and underdevelopment in the Third World and the ecological health of the planet.

The work products of the conference were also significant, Strong believed. While the Earth Charter was not achieved, the Rio Declaration, despite the weak wording, was "a profoundly important step forward" because it acknowledged such basic concepts as the "precautionary" and "polluter pays" principles, the need to change consumption patterns, and the importance of addressing global poverty. Agenda 21, he said, was "the most far-ranging, comprehensive action program ever approved by the international community. It was the first effort at a systemic integration of economic and ecological issues. Because it was constructed with inputs from virtually every sector of planetary society, from scientists to indigenous tribespeople, and because it was approved at the highest possible level of national government, Agenda 21 carries "a high degree of political authority," Strong said.

But the political success of the conference did not necessarily mean that it would produce immediate concrete results, Strong emphasized. In fact, he publicly questioned the commitment of national governments to follow through on the agreements they had reached in Rio. The most obvious evidence of that lack of commitment was the failure of the industrialized nations to come forward with financial resources to realize the goals of Agenda 21. "The industrialized countries, at a high level of generality, are willing to acknowledge the evidence that they are the source of most of these global risks. But they are unwilling to make any fundamental commitment to change the ways, the patterns of production and consumption, that have created those risks. . . . Rio did no produce that fundamental shift. But maybe it created the basis for that shift."

One of the major reasons for the failure to achieve that fundamental shift, Strong was convinced, was the absence of leadership in Rio. When a Brazilian journalist asked him how many world leaders were at the conference, he answered that he counted "about 120 presidents, prime ministers and assorted kings, but very few leaders."

While he was less than satisfied with the performance of Japan and the European Community, Strong was most disappointed with the failure of the United States to exercise leadership in Rio:

> The U.S. was a lagger rather than a leader. It was not only not a leader, it was almost a counter leader, resistant on some of the key issues. There is no question that the U.S. unwillingness to recognize any need for change in their own patterns of production and consumption had a pervasive political effect on other issues. . . . There's no question that the United States, as the strongest country on earth, was not recognized as the leading country. If the U.S. had combined its inherent strength with leadership, the results of the conference would obviously have been improved.

Nevertheless, Strong noted, even without powerful leadership, without a "driving engine," the summit achieved a "remarkable degree of agreement. Agenda 21, for all its weaknesses, is an amazing document." The other nations were not prepared to join the United States in sitting on the sidelines, he said.

Still, Strong conceded, the failure of U.S. leadership put the global commitment to carrying out the decisions of Rio very much in doubt.

. . .

Rio may wall have been a first look at an what the emerging post–Cold War global political structure will look like. The United States, unquestionably the sole remaining superpower, was not viewed as the world leader simply by default. It was not able to impose its will and block the international community from taking actions, such as the biodiversity treaty, that it opposed. It was not able to achieve goals, such as the forestry convention, that it pressed hard for. The United States often found itself at odds with Europe and Japan, its Cold War allies. And while the developing countries were not able to win the concessions they sought, their voices clearly carried greater weight than in previous international councils. The power gap between North and South, while still wide, appeared in Rio to be at least beginning to narrow. The reality of the need for cooperation to protect the global commons forced the industrialized countries to confront their dependence on the actions of poor nations for their future well-being.

As Pakistan's Jamsheed Marker pointed out after the conference, all the concessions made by the South for the protection of the global environment were "conditional upon cash resources being made available" by the North. Development and industrialization of the South cannot be stopped, and therefore, there must be transfers of cash and technology if it is to be done in an ecologically sound manner. A major achievement at the summit, he said, was "the universal acceptance that you cannot succeed in cleaning up the planet so long as poverty and underdevelopment exist." He added, however, that unbridled consumption by the rich countries is "the major disaster that is looming ahead of us."[30]

Ambassador Razali Ismail of Malaysia, who would become the first chairman of the new UN Commission on Sustainable Development, felt that much of the effort in Rio had been directed at trying to get the South to "shoulder the burden of change" needed to protect the global environment. The consumption patterns of the industrialized countries received relatively little attention, and "that is pretty unfair." The conference also produced few tangible changes such as the reduction of shipments of hazardous and nuclear wastes to the poor countries. Still, he said, it did produce Agenda 21, "the first major document we have that looks at the necessity of development in the North-South relationship. And in that document it says that $125 billion from the North will be necessary. And people are reminded by the document how far behind they are in meeting the official development aid targets set back in the 1960s."

In Rio, Malaysia was one of the most vocal opponents of international controls on forest exploitation and any other restrictions on the use of domestic resources, insisting that such restrictions infringed on its sovereign rights. "The Northern countries, having depleted their own resources, are now trying to globalize remaining resources, which means they are asking for a share of the resources that are in our back yard. That is very difficult to swallow," Razali said. But he conceded that "the old concept of absolute sovereignty is gone. You have satellites in the sky that can do remote sensing of our forests. The U.N. will now go into a country for humanitarian emergencies on occasion even if it does not have permission from the country to do so. It is pushing inward the old boundaries of sovereignty. And I think we have to accept this."[31]

But Prime Minister Gro Harlem Brundtland, whose World Commission report had been the inspiration for the Earth Summit, concluded that the unwillingness to move beyond narrow national interests was a major reason that Rio failed to produce a true "global partnership" to preserve the environment and move toward economic equity. What was missing in Rio, she said, was "a wider definition of democracy than national democracy. The global outlook, the feeling of democratic responsibility that carries across national borders, was not at absolute zero but it was largely missing." The Earth Summit, she said, had been perhaps one step on the "staircase" toward more democratic international governance. A large obstacle in the path toward that goal, she added, is that under the UN structure, decisions by the international community must be reached by consensus. That means that the process moves no faster than the slowest member of the community is willing to go. In a world threatened by accelerating ecological and economic decline, such a system is "not acceptable," she asserted. In the future, she said, decisions such as environmental treaties would have to incorporate supranationality as a principle and be binding rather than voluntary. And rather than requiring complete consensus, international standards could be reached by a "broad majority of nations" and backed by sanctions.

"We are not ready yet. I know that. But it will come, because global business cannot deal with different national rules for emissions and standards, with trade and product requirements, and because everything now crosses national borders from data systems to pollution. So democratic decisions have to be made to set limits and targets and to make social responsibility not only a part of a national agenda but a global one."[32]

. . .

The results of the conference drew mixed reviews from outside observers. Sir Crispin Tickell, the former British ambassador to the UN found that

> [i]n many ways the Rio Conference was a disappointment. Its final documents were badly drafted, stuffed with jargon, politically correct on the canons of UN-speech, shot through with ambiguities and lacing measurable commitment. Yet real progress was made: the Declaration of Principles was the best yet made and universally accepted. . . . Two legally binding conventions were signed . . . also a wishy-washy agreement on forestry. Agenda 21 [is] . . . a rag-bag [that] contain[s] the good and the bad, the valuable and the meretricious, but it represents hooks on which future agreements may come to be made."[33]

Lawrence E. Susskind, director of the M.I.T.–Harvard Public Disputes Program at the Harvard Law School felt that the "modest results" of the Earth Summit offered "conclusive evidence of the weaknesses of the existing environmental treaty-making system." The agreements reached in Rio, he wrote, represented "lowest common denominator results. The countries involved politicized the search for scientific understanding. . . . They minimized the search for creative options . . . They undervalued the importance of benefit sharing, focusing more on short-term economic costs than on long-term environment gains for future generations. . . . The final results were dominated by the most powerful states."[34] Norman Myers was disappointed that nations of the world had "argued and negotiated and produced piles of recommendations that reflected primarily the old order of individual nations with their individual interests. Hardly a single leader rose above parochial concerns to speak up for the Earth or the world."[35]

There were other disappointments to be found in Rio if one looked for them. Such bedrock problems as the effect of population growth, consumption, military activity, and trade on the environment were not given the attention they deserved and required. Scant sense of urgency was displayed about the need to change economically and ecologically self-defeating patterns of bilateral and multilateral development assistance, to create a more equitable system of international trade, and to ease the debt burden on the poor countries and reverse the net flow of capital from South to North. The insistence of developing countries that aid be given without conditions suggested that they did not necessarily plan to use financial and technical assistance to protect their domestic resource base. While the crucial role of business and industry was recognized, the summit did not attempt to spell out new norms of behavior to guide the transnational corporations, whose activities have such enormous impact on the global economy and ecology, toward becoming part of the foundation of sustainable development rather than obstacles.

"Sustainable development" was, of course, the mantra of the conference, repeatedly incanted by all participants. But the Rio conference failed, in the view of many observers, to weld environmental imperatives firmly to economic development in a systematic, programmatic way. Certainly, Agenda 21 set specific price tags for each environmental program and, in fact, contained a chapter on "integration of environment and development in decision-making." It also called for "capacity-building" to help create skills and institutions required by the developing countries to make use of the tools needed to attain economic critical mass and protect their environments. Missing from the output of the conference, however, was a broader vision of what it takes to build viable economies that would prosper over the long run within the carrying capacities of ecological systems at the local, regional, and global levels.[36]

. . .

To simply tally the failures of the Earth Summit, however, is to miss its potentially historic significance. The world may not have been saved in Rio, but it undoubtedly was changed. For the first time, the sovereign nations of the world gave formal recognition to the reality that the economic and ecological health of the planet are inextricably linked. In an address to the Economic and Social Council a month after the conference, Boutros Boutros-Ghali declared that "after Rio it is no longer credible to speak of the environment without putting it in to the context of economic and social development."[37]

Considering that it was created by a consensus of 178 diverse nations, Agenda 21, for all its weaknesses and mealy-mouthed wording, is a remarkable achievement, an unprecedented blueprint for global action to be taken during the rest of this century and into the next. The conference produced binding treaties to mitigate climate change and to preserve biological diversity, surprising additions to the body of international law considering that until a few years earlier few people in the world were even aware of the threat of global warming or of the rapid eradication of the planet's genetic resources.

But as Nitin Desai, the deputy secretary-general of the conference, frequently pointed out, Rio was as important as a *process* as it was for its products. It was, above all, an educative process in which the people of the world were forced to confront the knowledge that the way we as a species use the resources of the earth affects our ability to survive and prosper. It also drove home the lesson that unless we care for human needs and wants, we cannot preserve the habitability of the planet. National governments were asked to provide reports on the state of their own environmental efforts and plans for sustainable development, a process of induced introspection that inevitably prompted many capitals to take a fresh look at their environmental activities and were, for some, the basis of new national action plans.

The massive participation of environmentalists and other sectors of civil society in the Earth Summit process can only be regarded as long step forward in the democratization of international politics. Having been invited in the door, the NGOs served notice that they would not be willing to leave the room in future negotiations and that, indeed, they were prepared to act if governments would not. The integration of the independent sector into the Rio process also may have contributed to opening up the politics of some national governments, particularly in those parts of the world where there is little tradition of participatory democracy.

Perhaps the greatest significance of Rio was the potential it presented for creating a new basis of international relations. The old system of collective security, built on ideological competition, armed confrontation, and the fear of mutual destruction, had recently evaporated. Now the imminent threats to security were recognized to be the twinned crises of economic inequity and ecological destruction. Peace, it was acknowledged in the rhetoric of world leaders in Rio, had a new name and a new face: a *Pax Gaia*, to be built, as the ecotheologist Thomas Berry had taught, on the understanding "that the earth is a single community composed of all its geological, biological and human components."[38] Collective security would henceforward be based on cooperation among the people, governments, and institutions of the world to secure the safety of the planet and the well-being of its inhabitants into the next millennium.

That, at any rate, was the message from the summit.

Rio was a doorway opening to a new post–Cold War, postindustrial era—an era of more humane economics and benign technologies, greater concern for the vulnerability of Mother Earth and her creatures, and growing equity among nations and between generations. Just possibly, the new laws, principles, promises, prescriptions, and goals set forth in Rio laid the foundation for a second enlightenment—a green enlightenment.

But the nations and peoples of the world would still have to muster the will and the energy to step across the threshold of that open doorway. They did not do so in Rio. The Earth Summit was chiefly about ideas, visions, aspirations, hopes. After Brazil, actions would be required.

Chapter 12

After Brazil

When you come to a fork in the road, take it.
—Attributed to Yogi Berra

After the summit the interest of the media and the international public in the issues discussed there vanished quickly—it seemed almost instantaneously, as if a switch had been turned off. With public pressures removed, most national leaders and governments, particularly in the North, shoved sustainable development toward the bottom of their agendas. The attention of the international community shifted quickly from long-range plans to protect the environment and promote economic development to customary economic self-interest and to the immediate necessity of trying to stop large numbers of people from murdering each other in various parts of the world.

It was back to business as usual.

. . .

That is not to say that the goals set at the Earth Summit were abandoned after Brazil. On the contrary, there was a substantial amount of follow-up, starting with the UN itself and proceeding down through national governments, regional and local authorities, and the private sector.

Within six months after Rio the General Assembly of the UN had enacted the key institutional changes agreed on at the summit:

It created a new UN Department for Policy Coordination and Sustainable Development to coordinate economic and environmental activities within the UN system. Nitin Desai, Maurice Strong's deputy for the Rio conference, was named under-secretary of the UN in charge of the new department.

It authorized a new intergovernmental, fifty-three-member, ministerial-level Commission on Sustainable Development to monitor the progress—or lack of it—by national governments and international organizations in carrying out Agenda 21. Ambassador Razali Ismail of Malaysia was named as the commission's first chairman, and Desai was put in charge of its management. The commission reports to the secretary-general of the UN through the Economic and Social Council (ECOSOC), an arrangement that failed to please many environmental activists because of the council's reputation for inaction and ideological squabbling. But supporters of the plan pointed out that the Commission on Human Rights, which followed the same arrangement, had not been inhibited. Some veteran UN watchers also felt that the new responsibilities might help revitalize ECOSOC.

In a move not directly related to Rio but that over the long run may be vital to achieving its sweeping goals, a group of eminent persons from around the world created the Independent Commission on Global Governance. Its mission, approved although not formally mandated by the UN, is to review the way the international community currently attempts to govern itself and suggest ways to create an improved system of international cooperation for peace, prosperity, and democracy. After its first meeting in Geneva in February 1993, the co-chairmen of the commission, former prime minister Ingvar Carlsson of Sweden and former Commonwealth secretary-general Sir Shridath Ramphal, said that it would look at reforms for the UN and its specialized agencies, including the World Bank and other Bretton Woods institutions and the General Agreement on Tariffs and Trade (GATT) and would make recommendations on issues such as security and disarmament, human rights, and conflict intervention, as well as environment and development.[1]

The establishment of the commission was tacit acknowledgment by the UN and its member nations that the current system of international governance—if indeed it can even be called that—is not capable of adequately addressing those issues.

The international community also moved forward on one major front to begin to meet the financial needs of sustainable development when governments agreed to change the Global Environment Facility (GEF) from a pilot project to a interim and perhaps permanent body and by opening its management to broader participation. The GEF (pronounced "Jeff"), is a joint project of the World Bank, the UN Environment Program, and the UN Development Program, formed in 1991 to provide grants to developing countries to help them deal with climate change, biological diversity, pollution of international waters, and depletion of the ozone layer.[2] Although the developing countries had sought a new, independent financial mechanism to fund money for environmental projects because of unhappiness with the World Bank, in Rio it had been decided instead to restructure GEF as the key international financial channel of money for sustainable development.

Funding for the "new" GEF, which was headed by Mohammed El-Ashry, who had been environment director of the World Bank and before that a vice president of the World Resources Institute, was raised from less than $800 million to $2 billion for a three-year period. Its mandate was expanded to include projects to combat land degradation, its governance was expanded to included representatives of developing countries on its governing council, and its meetings would be open to nongovernmental observers. In a further effort to democratize the process, many of the projects of the GEF would be proposed at the community level and would not have to be approved by national governments before receiving funding. The facility, El-Ashry said, "can help with the badly needed integration of environment and development on the ground. We are an important first step. But let's not fool ourselves. We were a compromise. A lot more is needed."[3]

Another meaningful step toward integrating the UN approach to environment and development was taken in 1993, when James Gustave (Gus) Speth of the United States was named head of the UN Development Program. Speth, a lifelong environmentalist who had been president of the World Resources Institute and chairman of President Jimmy Carter's Council on Environmental Quality, promptly set his agency the mission of abetting "sustainable human development" in the countries it serves. Other agencies, including the UN Environment Program, restructured their priorities to reflect the decisions reached at the summit. In the spirit of Rio, the UN agencies, famous for their internecine bickering and competition, continued to cooperate on matters pertaining to sustainable development.

By the end of 1994, Nitin Desai was able to point to a substantial number of other advances by the international community along the front opened at Rio.[4]

- Under the leadership of Sweden's Bo Kjellen, member nations of the UN negotiated a convention to join in combatting the spread of deserts, a treaty sought particularly by African countries, that included binding commitments integrating environment and development. National leaders at the summit had pledged to support such a treaty in the future, but the rapidity with which it was negotiated surprised many.
- The environmental and economic problems of small island nations were addressed at a special UN conference, which produced a compact for protecting the fragile environments of those ocean-girt countries.
- Before 1994 ended, 104 nations had ratified the climate change treaty.
- The industrialized countries agreed to ban the export of hazardous wastes to the developing countries, a significant advance on concessions made in Rio.
- At a meeting in April 1994, governments agreed to establish an international forum on chemical safety.

One of the most meaningful post-Rio activities of the world community was the International Conference on Population and Development, held in Cairo in September 1994. Despite opposition to family planning and abortion by the

Vatican and several fundamentalist Moslem countries, broad consensus was reached among the 180 participating nations to adopt policies that would stabilize global population over the next twenty years. While the policies would include family planning, they also stressed the importance of public health, education, human welfare, and especially the rights of women. The Cairo population plan called for spending some $17 billion annually on family planning. No specific numerical targets were set by Cairo program, but it was estimated that if it were adopted, world population, currently 5.67 billion, would stabilize at 7.27 billion by 2015. If the program failed, population would rise to 7.921 billion in 2015 and to 12.5 billion in 2030. Even the Vatican, which had shunned previous population agreements, agreed to part of the Cairo accord because of concern it was being "isolated" in the population debate.[5]

The Cairo conference did not go as far as some population stabilization advocates had been hoping, and some critics complained that it concentrated on issues of women's empowerment while paying too little attention to environmental threats. But there is little question that it took this bedrock issue of environmental and economic sustainability a major step beyond agreements reached in Rio.

A number of other UN gatherings linked to the sustainable development goal were scheduled over the next few years, including a World Summit for Social Development in Copenhagen in 1995; the Fourth World Conference on Women in Beijing; a Habitat II conference on human shelter, in Istanbul; as well as conferences on migrants and public administration. One veteran UN official expressed concern that, because of the frequency of UN-sponsored conferences, the issues raised at them tended to disappear into a "black hole" of oblivion and trivialized the significance of the Earth Summit. But Nitin Desai, among others, found that those follow-up conferences were part of a reinvigorated search by the community of nations for a new consensus on global cooperation.[6]

Some scholars describe the Rio conference as the opening of an entirely new chapter of international law. Peter Sand, a legal adviser to the World Bank who had served in a similar role on Strong's secretariat, found "a major paradigm shift, in Rio from international 'environmental law' to a new (and as yet to be defined) 'law of sustainable development.' "[7] Sand recalled that during the preparations for the summit, Brazilian delegate Pedro Motta Pinto Coelho persuaded fellow negotiators to change the wording from "environmental law" to "international law of sustainable development" throughout the text of Agenda 21. Then he said: "That will keep you lawyers busy well into the 21st century."[8]

Of the legal documents produced in Rio, the treaties on climate change and preserving biological diversity, because they are binding on signatories, are what are traditionally considered to be "hard law," while the nonbinding Rio

Declaration and Agenda 21 are regarded as "soft law." But as Sand noted, the two conventions were only framework treaties without real teeth. They could not do much to mitigate global warming or to create a new "global regime" to preserve the earth's living resources unless and until protocols were added to require specific actions of national governments. On the other hand, the history of international law suggested that nonbinding principles such as the polluter-pays principle, common but differentiated responsibilities among poor and rich countries for sustainable development, and the precautionary principle spelled out in the "soft" Rio Declaration have a tendency to "harden" fairly rapidly.[9]

. . .

Sporadic activity by national governments in the months following June 1992 suggested at least a degree of readiness to move forward with the agreements of the Earth Summit. By the middle of 1994, fully a third of all governments had set up some sort of national sustainable development commission or similar body. Virtually all countries had prepared national reports outlining their environment and development leads. Several countries—including, remarkably, China—had drafted their own Agenda 21, translating the proposals presented at Rio into programs tailored to their own national problems and needs. India and the Philippines also launched strong post-Rio initiatives. The Central American nations joined in an "alliance for sustainable development." To Maurice Strong, the response was "surprisingly better among the developing countries than among the developed countries."[10]

Perhaps the important change in a national capital in the post-Rio period took place in Washington, D.C., when Bill Clinton was elected president and Al Gore, the chief herald of global environmental protection in the U.S. Congress, became his vice president. Strong regarded the election as a potential "sea change" at the center of power in America that could produce a change in attitudes and commitments to environment and development. For a time his optimism seemed justified. President Clinton announced that the United States would sign the biodiversity treaty and would agree to the goals and timetables for reducing carbon dioxide emissions that had been rejected by his predecessor. As part of fulfilling the nation's obligations under the biodiversity treaty, Clinton authorized the first comprehensive survey of biological resources in the United States. Timothy Wirth, who had been Gore's ally on environmental issues in the Senate, was named under-secretary of state for environment, science, and related affairs. Under Wirth's aggressive prodding, the United States resumed its leadership role on population issues, a role it had abandoned during the Reagan presidency. The appointment of Gus Speth as head of the UN Development Program, traditionally a post reserved for an American, was a direct result of the changed administration in Washington.

Clinton also created the President's Council on Sustainable Development to advise him on how the United States should respond to the decisions taken in Rio. He named Jonathan Lash, who succeeded Speth as head of the World Resources Institute, and David Buzzelli, vice president for environment of the Dow Chemical Company, as co-chairmen of the council. Some U.S. environmentalists felt that the work of the council was too slow and diffuse. But by the end of 1994, halfway through the council's mandated life, Buzzelli felt it was positioning itself to offer important policy recommendations in key areas such as agriculture, industry, energy, and the "exciting" field of sustainable communities. "Rio," he said, "brought on an awareness of the need for sustainable development that we just never had before." This is particularly true among corporations, which, he believes, are beginning to do things very differently, including reducing waste on a voluntary basis and teaching their customers how to conserve materials.[11] Buzzelli's boss, Dow CEO Frank Popoff, declared in a speech to chemical industry executives: "Some in the chemical industry believe that the only benefit of improving our environmental performance is a better public image. But I believe that sustainable development is a matter of economic survival."[12]

Responses to Rio also were visible on the state, provincial, and municipal levels around the world. Local communities in India, Canada, and Sweden adopted their own sustainable development programs. The International Union of Local Authorities formally adopted Agenda 21. A number of cities in the United States launched initiatives based on the Rio agenda, including "Sustainable Seattle." The governor of Kentucky convened a conference in 1993 to plan ways the states could move forward on sustainable development. Individual states, including New York, held their own meetings on the issue. Iowa produced the "Earth Charter," an environmental platform based on the lessons produced at the Earth Summit.[13]

Environmental groups and other civil sector organizations that had become engaged in the Rio process remained active after Brazil. Many of them established contact with the new Earth Council, a private organization of persons prominent in the sustainable development debate; it was founded by Maurice Strong and based in Costa Rica. The council took it upon itself to monitor and assess the follow-through to Rio, to serve as an interchange of information on sustainable development activities, and to be a kind of global switchboard for communication among nongovernmental organizations (NGOs) around the world. By the end of 1994 the council had established communication links with some twenty-five thousand NGOs.[14] Alicea Barcena, a former undersecretary of ecology in the Mexican government who had been a valued member of Strong's Rio secretariat, was named as its first executive director. "The idea," explained Barcena, "is to have a totally independent body that gives equal status to all sectors, that provides a forum where all sectors can participate. We want to be helpful to governments but governments should not be the only actors."[15]

Many of the environmentalist networks formed in the years before Rio, such as the Climate Change Network, the Multilateral Development Bank Campaign, and the Pesticide Action Network, had expanded greatly during the Earth Summit process as activists, particularly those from developing countries who had never before participated in international efforts, learned the tactics of putting pressure on governments and international governments and joined in the fray. Some 1,400 NGOs were accredited to make representations to the new Commission on Sustainable Development. One problem, however, was that financial resources that been made available to allow Third World environmentalists to participate in the Rio process had largely dried up and thus the activists' input was limited to commission deliberations in New York. Nevertheless, NGOs kept the pressure on governments by information gathering, lobbying, and public exposure of inaction, to try to get them to meet the commitments they had signed in Rio.

One of the most important legacies of the Earth Summit is the sense of community, of common goals, that developed among participating environmentalists from around the world. While there was still a residue of distrust and suspicion among some nongovernmental activists from the Third World, particularly among those concerned primarily with development as opposed to the environment, the wide breach that had been so apparent during the preparatory phase had been largely closed. As one U.S. environmental leader put it, "We became comfortable with each other and learned to trust one another."[16]

. . .

A series of dramatic political changes on the world stage also seemed to offer favorable omens for a new era of harmony and stability in which the much talked about new world order could be constructed. In one of the most unexpected and heartening manifestations of the expansion of democracy that has transformed the world in the last years of the twentieth century, apartheid was at last eliminated and a black majority government was installed in South Africa. A seemingly miraculous, if fragile, peace agreement was reached between Israel and the Palestine Liberation Organization. An equally miraculous and fragile cease-fire was declared in the long war of terror in Northern Ireland. A democratically elected regime was restored in the tortured island nation of Haiti. Threatened anarchy was averted in the Russian republic. The unification of western Europe proceeded despite the many hurdles created by doubts about sharing sovereign powers. North Korea seemingly agreed to yield its nuclear weapons capacity, easing tensions on the always explosive Korean peninsula. A strong surge in China's economy offered hope that democratic reforms might follow in that vast, enigmatic country. Mexico joined the United States and Canada in a North American Free Trade Agreement (NAFTA) which included promises of environmental as well as economic cooperation. A successful conclusion to the Uruguay Round of GATT negotiations held out the possibility that freer world trade would spur greater international cooperation and integration.

. . .

So much for the good news.

As Maurice Strong noted two-and-a-half years after Rio, the "fundamental changes" that had been called for at the summit had not taken place.[17] (Strong himself had accepted an invitation to become chairman of Ontario Hydro, North America's largest utility, in order, he said, to fight for sustainable development in "the trenches" of the corporate sector.) The global environment was still in danger. Poverty was still rife and economic takeoff just as elusive in most of the developing world. The voracious consumption of resources and production of waste in the rich developed countries continued more or less unabated. Signatories of the climate change treaty failed to agree on specific reductions of greenhouse gases, and projections showed that carbon dioxide emissions would begin rising even more sharply after 2000. At the end of 1994, members of the European Union indefinitely shelved a plan to impose a common energy tax as a means of reducing carbon dioxide emissions.[18] The world's flora and fauna continued to dwindle as habitat destruction went on apace, particularly in the species-rich tropics, and consumption of such resources as fish stocks continued to outstrip their ability to replace themselves. Most of the industrialized nations did not even bother to send ministerial-level delegates to a conference of parties to the biodiversity treaty held in Geneva at the end of 1994.[19] Plans to follow up the forest principles agreed to in Rio with an international law to preserve the planet's forest appeared to have been discarded in the diplomatic dead letter box, despite the formation of a private international commission to study the forest issue. Population growth, along with poverty, hunger, and war, continued to spur mass migrations and did much to destabilize societies around the world.

Ominously, much of the political will to safeguard the planet that had been signaled by the spirited rhetoric of Rio seemed to have melted away. The international community, or at least the industrialized North, appeared to have gone into full retreat on many of the commitments made in at the Earth Summit.

"Remember the environment?" plaintively asked John Stackhouse, development issues reporter for the *Toronto Globe and Star,* in an article published at the end of 1994. "More than two years after the United Nations brought together world leaders in Rio de Janeiro to talk about a cleaner, greener planet and a fuzzy concept called sustainable development, environmental issues are showing signs of becoming an endangered species on the international-development agenda."[20]

In fact, however, whatever progress was made in meeting the promises of Rio—precious little to be sure—was almost entirely on the environmental side of the global bargain, just as the poorer countries had feared. Commitments made at the summit to restimulate the development process with new and additional development funding went largely unhonored. Not only did the rich countries fail to make good on the modest amounts of new and additional money for development they had pledged in Brazil (with a few honorable exceptions such as the Netherlands, Denmark, Sweden, and Norway), official development assistance actually *declined* by 10 percent in the two years after Rio. And it was clear that the shrinking economic pie would have to be cut into more pieces to pay for the programs agreed to at the population conference in Cairo and for new obligations that arose from social development and women's conferences and other UN-sponsored gatherings in the offing. Promises of technical assistance, made in Rio to developing countries, also were largely unkept.

By October 1994 international officials and other eminent persons, gathered in Tokyo for the Conference on Global Environmental Action concluded in their final declaration that "the financial resources available for achieving sustainable development do not adequately reflect the sense of crisis that we face."[21]

The UN itself was facing a "financial crisis" according to Secretary Boutros Boutros-Ghali.[22] Strong warned that even the future of the UN Environment Program, which he had founded in 1974, was now in doubt.[23]

Momentum flagged on many fronts. While the Commission on Sustainable Development was up and running under capable hands, it seemed to be evolving, according to one veteran UN official, into "a political club for intellectual discussion among ministers and diplomats. It has inadequate resources and no teeth."[24] One of the reasons for this evolution, the official said, was that the commission and the issues with which it is engaged were receiving virtually no attention from heads of government after Rio. Many of the sustainable development commissions and councils set up by national governments after Rio, like the one created by President Clinton, were solely consultative or advisory bodies, with no operational responsibilities and with limited ability to influence basic governmental policy-making.

. . .

Why had the flame of sustainable development, which burned so brightly in Rio, receded to a dim flicker in only two years? One reason, suggested by Kamal Nath, the environment minister of India, was that "the attention of world leaders seems to have moved to other international issues. The agenda with environment at the top seems to have changed to one of conflict resolution."[25]

Certainly peacekeeping and crisis management missions over the past few years have strained the resources of the UN and made additional demands on many of the governments that provide the bulk of financial support for the international organization. Attempting to deal with famine and clan warfare in Somalia, tribal genocide and mass migration in Rwanda, continuing civil war in Angola, and the murderous ethnic conflict Bosnia; trying to help the Russian republic and its former satellites switch economic systems; and attempting to cool reemerging nationalistic, ethnic, and religious strife that is creating flash points on a global map seemed to overtax the energy of the UN and its member states.

At the same time, a number of the industrialized nations were slow in emerging from a long recession and were experiencing unaccustomed levels of unemployment and reductions in social services. Not surprisingly, the slackening of domestic economic vitality was accompanied by what has been dubbed "aid fatigue"—unwillingness to send money from depleted treasuries to developing countries, many if not most of which had not seemed able to put past billions in development assistance to effective use.

But the agreements reached in Rio were also self-interested. What, after all, could be more in the interest of governments and the constituents they serve than preserving the earth's life-support systems? What economic priority could be higher than reducing global economic inequities that were not only putting unsustainable pressures on those systems but would almost certainly lead to dangerous geopolitical instability if unaddressed? The failure to keep the promises made in Rio, therefore, must spring not only from immediate crises but from flaws in the structural capacity of the human community to manage its most pressing affairs.

For one thing, national and international institutions quite obviously have yet to adjust to the sudden end of the East-West confrontation. For nearly fifty years the Cold War had imposed an understandable if cruel set of rules governing the behavior of nation-states and their international associations. Nations generally understood where they stood in relation to the rest of the global community. Diplomatic waters were dangerous but navigable. The United States and the Soviet Union were the dominant powers, but when they exceeded the limits of their power, as in Vietnam and Afghanistan, they were forced to absorb costly and painful lessons. Alliances were stable, and the nonaligned nations learned how to use the pervasive ideological competition to their own advantage. The UN served as a emergency rescue squad, a debating society, and a provider of global social services but also, it must be said, as a tantalizing promise that the world could, somehow, learn to govern itself.

With the cessation of the Cold War, however, the international polity was set adrift without charts in an unfamiliar sea. The end of biopolarity set the needle of geopolitics into a bewildered spin. The collapse of the Soviet Union and the end of the Warsaw Pact was being followed, it appeared at the beginning of 1995, by the slow disintegration of the Western Alliance. The old allies seemed to be casting about for new diplomatic bearings, with western Europe turning its eyes to the East and the United States to the South of its own hemisphere. And as Barbara Crossette of the *New York Times* observed, "the fraternal third world" that charismatic leaders of the 1950s and 1960s—Jawaharlal Nehru, Gamal Abdel Nasser, Kwame Nkrumah, Sukarno, and Zhou Enlai—had hoped to build, was effectively "dead" by 1994.[26] Despite efforts to maintain solidarity, the once more or less solid South was split between increasingly poor and increasingly affluent countries, between nations with democratic and authoritarian regimes. The old "second world" of the Soviet Union and its Socialist allies also had disintegrated, and its fragments were now competing with its former clients in the nonaligned world for the crumbs of economic assistance falling from the tables of the European Community, North America, and Japan.

In those countries of the developing world that were achieving economic takeoff, primarily in parts of Asia and Latin America, there was a clear danger that patterns of production and consumption would replicate the wasteful, unsustainable practices of the North and multiply the environmentally destructive effects of industrialism to levels that could not be sustained by the earth's life-support system.

In theory, the UN could be a beacon, leading the international community through these perplexing waters to the safe haven of a new system of political stability. And as noted in the previous chapter, the summit meeting on environment and development was a major initiative by the UN system to lay the foundation of a rebuilt structure of collective security based on cooperation to protect the planet and the well-being of all of its inhabitants now and in future generations. A livable planet and economic and social justice for all of its people is not a bad start on a new central organizing principle.

Events after Brazil have demonstrated, however, that the UN is much better in laying out organizing principles than as an organizer of action by nations and peoples. The sluggishness and ineffectiveness of its response to recent crises, particularly its long inability to end the tragic fiasco in Bosnia, has once again confirmed that the UN does not have the power or capacity by itself to determine the course of geopolitics. It cannot by itself, therefore, assure that the principles and pledges made in Rio are observed by its member nations.

As former Under-Secretary-General Brian Urquhart observed, "The U.N. is a channel for leadership. It cannot exercise leadership. You can't move this international body without Olympian leadership. . . . The United States provided remarkable leadership after World War II but it no longer does so."[27]

Urquhart made those remarks during the preparatory process for the Rio meeting, when the Bush administration was deliberately abdicating leadership for political reasons. But the "sea change" in U.S. leadership that Strong and others had hoped for after the election of Bill Clinton and Al Gore also failed to materialize, at least in the first two years of their administration. Clinton's weak presidency was overwhelmed by fierce political and ideological opposition conducted by the Republican minority. The Republicans, joined by conservative Democrats, were able to block most of his domestic agenda, including virtually all of his efforts to improve antipollution laws and protect public lands and endangered species. They provided no help in strengthening an uncertain foreign policy, except for supporting the NAFTA and GATT treaties, which were favored by most American corporations, and actively sought to undermine many of the administration's diplomatic initiatives.

The United States, the only superpower left standing in the geopolitical coliseum as the end of the twentieth century approached, was a confused, bumbling giant, unsure of where to turn, unable to exercise its strength. Despite efforts by Vice President Gore, Tim Wirth at the State Department, and others in the administration, Washington proved incapable of leading the international community toward a new system of collective security based on the principles enunciated in Rio.

With the smashing triumph of right-wing politics in the U.S. Congressional elections of 1994, any chance that America would exercise such leadership in the near future was extinguished. It was clear that not only would the reactionary Republican majority fail to support major new environmental and development assistance programs, it was prepared to conduct a full-scale withdrawal. Increased development assistance? Senator Jesse Helms of North Carolina, who would be the next chairman of the Senate Foreign Relations Committee, said after the election that he distrusted "cutesy" ideas such as sustainable development; Senator Mitch McConnell, who would head a key subcommittee of the Senate Appropriations Committee, said he thought the U.S. Agency for International Development (AID) should be eliminated.[28] The only increase in spending on the nation's global interests that was likely to be approved by the new majority would be to resume building up the defense budget. It was also clear that the Republicans would seek to roll back many domestic environmental and resource conservation programs as part of their effort to reduce the role of government and give capital and corporations freer rein to pollute and to exploit resources and consumers without fear of penalty.

. . .

At the middle of the last decade of the century, therefore, conditions did not seem propitious for a global transition to sustainable development. As Maurice Strong accurately described it, "We are in a dangerous and frustrating interregnum in which political leaders are more prone to resort to *ad hoc* responses to the problems and pressures they face based on considerations of narrow, short term political expediency than to face up to the root causes of these problems."[29]

Instead of a new relationship between humanity and the planet it dwells on, instead of a new relationship of equity among peoples and between generations, the international community seemed to be headed to a period of social Darwinism based on individualism, nationalism, ethnicity, and economic competition. The events in Bosnia, the Caucasus, Somalia, and Rwanda suggest that humans are as likely to retribalize as to unite as a species to build a *Pax Gaia.* The current politics of the rich industrial powers, again with a few honorable exceptions, offers little immediate hope that they will take decisive action to reduce their own excessive consumption or waste, to halt the degradation of the environment, or to keep the gap between the poor and the rich from continuing to widen.

. . .

And yet those problems will have to be addressed; the transition to sustainable development must be made. The alternative, as many commentators have warned, is a civilizational crisis, as the physical, biological, political, social, and economic systems that sustain us break down.

But how, given the failures and inadequacies of national governments and international institutions, could the transition be started? Where could it happen? Who will do it?

One answer is that it *has* started. It is happening in cities, local communities, and rural areas in all parts of the world. It is being done by community and grass-roots groups, by environmental organizations, by businesses, by private foundations, by individuals. Sometimes the effort is being aided by governments; more often not. But with or without governments, projects aimed at environmentally sustainable development are sprouting from native soil around the globe.

Chapter 13

Doing It

Be a gardener.
Dig a ditch,
toil and sweat,
and turn the earth upside down
and seek the deepness
and water the plants in time.
Continue this labor
and make sweet floods to run
and noble and abundant fruits
to spring.
—Julian of Norwich

Although it sits squarely astride the Equator, the dusty, bustling town of Nanyuki is enjoying cool, springlike weather. The temperature on this sunny afternoon in May 1993 is in the invigorating mid-seventies, reflecting the fact that the town is well up the lower slope of Mount Kenya, which rises over it to snow-covered crags nearly eighteen thousand feet above sea level. Five of us—Joseph Ndungu and Martin Kamau, field directors of the Laikipia project; my wife and I; and the driver—pack ourselves into a small, ancient pickup truck, which obviously has lost all memory of shock absorbers, and go bouncing jarringly along a deeply rutted dirt road up a gentle gradient of the great mountain.

After some forty-five spine-wrenching minutes, the pickup pulls to the side of the road, and we work our way out the doors. Craning our heads back, we see the peak far above us, half covered by a black rain cloud, half a dazzling white of sunlight reflected off the snow. Turning our backs to the mountain, we look down at small fields of alternating green and yellow; tiny, circular mud huts covered with thatch that house Kikuyu families; and scattered clusters of cattle and goats. Beyond are more fields and huts and still more below those down to a distant valley floor obscured by haze. Far off in the distance are more high mountains.

We are standing, Ndungu and Kamau explain, in the middle of an ecological system that shifts from the relatively moist upper slopes of the mountain to a dry savanna below. Before the arrival of the British what is now Laikipia Province was occupied by Masai herders. When Kenya was colonized, the Masai were driven out, and Laikipia was transformed into what Ndungu called a "white island" of big cattle ranches whose earnings made a few Englishmen wealthy. But after independence, when the whites departed, the Masai were not permitted to return to their ancestral pastureland. Instead, the land was given by the new government to the Kikuyu, the politically dominant tribal group in Kenya. Kikuyu farmers were moved from the overcrowded Central Province to Laikipia's newly vacated pastureland.

At first, each farmer was able to work fairly large holdings, sometimes reaching as many as fifty acres. But developers, working hand-in-hand with corrupt government officials, repeatedly subdivided the land, and now the average farm is no more than two to five acres. That amount of acreage is too little to support even a nuclear Kenyan family, and such families tend to be large. Moreover, the Kikuyu farmers were used to the agricultural practices of the rainy Central Province and did not know how best to cultivate the land in the semidry conditions of their new home. The result was a downward spiral in the condition and productivity of the land and the quality of life for its human inhabitants. The soil, relatively fertile although dry, in many areas became compacted to the hardness of concrete from severe overgrazing. Water, overused for irrigation on the upslope portion of the ecosystem, became increasingly unavailable downslope. Vegetation grew thinner, soil eroded, and wildlife became scarce, particularly the elephant herds that attracted tourists and much needed cash to the area. Because the tiny farms were no longer able to support bare subsistence for even the smaller nuclear families, many of the men left Laikipia to look for work in Nairobi, the great metropolis of East Africa, some five hours to the south by bus. Some of the men found jobs and are sending home regular stipends to their families. Others simply disappeared into densely crowded, crime-, drug-, and disease-infested city. For those who remained, especially the women and children, life was increasingly hard.

Then, in 1984 Nairobi University joined with Switzerland's Bern University to create a program to rescue Laikipia's economy and the ecological base on which it rests. The chief thrust of the Laikipia project was to find the optimum sustainable use for all parts the land by careful study of each of the subecosystems. For example, computers aided an examination of the grass and water resources and were able to provide figures for the maximum number of cattle that could be grazed on each pasture on a permanent basis. To ease the pressure on the rapidly disappearing forests, alternatives were suggested for cooking fuel, such as burning sawdust from the local mills. Minimum tillage farming was encouraged to reduce erosion. As Kamau noted, the farmers were too poor, in any case, to afford pesticides and fertilizers. Project workers also introduced agriforestry, a process involving the interplanting of trees and food crops to help conserve the soil and water and to provide wood and other benefits of trees. Catchment areas were built to store water that would otherwise run off the mountain. A ration of water is reserved for the elephant herds to assure their continued survival as a tourist attraction. The project also is working to upgrade social services, including schools, medical clinics, and family planning assistance.

Much of the technical work on the program is done by students and academics, many from Nairobi University. The director of the project is Hanspeter Lineger, a Swiss ecologist and geographer from Bern University. But all of the basic decisions on land and water use and infrastructure investments, as well as on social services, are made by the people of Laikipia themselves. More than 360 local councils have been formed to discuss and oversee all of the project's activities. Because so many men are away in Nairobi, the councils are chiefly run by women, who have become the chief economic decision makers. As Ndungu and Kamau underscored, this is an almost revolutionary transformation for Kenyan society.

After ten years, the Laikipia project is producing modest but real results. Crop yields are up although still only about 40 percent of maximum capacity. Soil erosion is slowing. Many participants in the project are better fed, healthier, and by all accounts, happier. The quality of their lives is improving.

At one farmstead a galvanized roof has been put on top of an expanded and whitewashed mud cottage. Rainwater falling on the roof is caught in a cistern. Neat rows of beans and corn are greening beneath the branches of equally neat rows of young, fast-growing grivillea trees imported from Australia. The side roots of the trees have been pruned so that they draw water only from their deep tap roots and leave the water close the surface available for the food crops. The trees help hold the soil and water and will provide firewood for the farmstead. Near the cottage a young boy tends two busily grazing goats. The pastoral scene projects a aura of simple but unmistakable prosperity that contrasts dramatically with the deepening rural poverty of much of Kenya.

"This is very different from traditional development projects, which depend on high inputs of capital and technical assistance," Liniger explained. "Here we make much more use of ground up capacity. We have worked with the local farmers from the beginning. We have had more trouble convincing the government of the need for changes in land use." The Swiss geographer added, however, that while only small amounts of outside funding were necessary for the project, it needed to be long-term funding. "The program will work over time. We need another ten years. But donors change their interests and their programs every ten years. Who is ready for sustainable funding?"[1]

. . .

Over the past decade or so, Africa has been regarded as a textbook on how the development process can fail. Despite large inflows of capital and technical assistance, poverty continued to rise and degradation of the land and its resources, particularly its wildlife resources, accelerated. Hunger and disease continued to widen. In countries such as Kenya corruption and runaway population growth helped cut the ground out from under the development process, along with an excessive dependence on export crops such as coffee and cocoa, which yielded sharply declining earnings in the face of depressed prices on the world market.

Mismanagement also undercut efforts of countries like Kenya to achieve economic takeoff, even in the management of its tourist trade. In 1993, I revisited the great Masai Mara preserve along the border with Tanzania after an eleven-year absence and was appalled by the devastating impact that excessive tourist traffic had upon the land and its wildlife. The gently rolling savanna, which had seemed on my earlier visit to be what the Garden of Eden must have looked like, now had been turned in many places into a moonscape, with the soil rutted and torn up by the endless crisscrossing of four-wheel-drive vehicles loaded with camera-toting tourists. The wildlife seemed much scarcer than I recalled. A decade earlier, lions were to be seen in abundance. On this trip our driver spent the whole morning searching, until we came upon a solitary, stately male lion—already surrounded by five other Land Rovers. The high-volume, relatively low-cost tourist trade encouraged by Kenyan policy does bring in substantial amounts of cash. But as Joseph Lynam, a Nairobi-based senior agricultural scientist for the Rockefeller Foundation, pointed out, most of the hotels, tour companies, and transportation companies that receive that cash are not Kenyan but foreign-owned, and the money soon leaves the country. That is certainly not a recipe for sustainable development.

"The outlook for Africa is bleak," Lynam contended. "There is little capacity in urban areas for industrialization. Southeast Asia is getting all the exported jobs from the industrialized countries. The agricultural area is depressed, and at the same time, population is growing rapidly. Institutions are generally weak. The international community is withdrawing its aid, particularly in Kenya because of the corruption. So how do you plan for sustainable development in Africa?"[2] Roger D. Stone, in his excellent book *The Nature of Development,* suggested a number of other formidable obstacles to development in Africa. These include generally hostile natural environments, particularly poor soils; decimating epidemics and widespread ill-health; and a history in modern times of domination by the North that took the form of slavery, colonialism, and in recent times, an economic system, particularly an international trade regime, imposed to satisfy the needs of Europe rather than of the Africans. Added to those problems are the corrupt governments, authoritarian regimes, lack of administrative capacity, tribal conflict, spreading crime and anarchy, and other internal political problems afflicting a number of African nations. "Nor," wrote Stone, "did the aid agencies help much. The voluminous literature on the subject is replete with examples of failed dam projects and irrigation schemes, of integrated rural development programs that produced neither integration nor development, of disease and environmental degradation arriving along with aid-donor largesse."[3]

And yet Africa is not a monochrome of failed development. Across the continent there are examples, such as the Laikipia project, of grass-roots developments that offer hope, not only of alleviating hunger and poverty but also that the land and the life it nourishes will be preserved. Sometimes these projects are assisted and guided by national governments or multinational agencies. More often they are helped by indigenous or international nongovernmental organizations (NGOs). Usually the projects require some form of outside financial and technical assistance, particularly to help build the skills of local participants and to give them the tools they need for the job. But virtually without exception, the development projects that seem to work are those that are built from the bottom up, with communities and individual citizens doing the planning, supplying the ideas, and making decisions.

"Governments don't empower people with information and skills," asserted Gilbert Arum of Kengo, an alliance of Kenyan NGOs. "Outside experts only come in and create more problems. Africans have to get out of this themselves." Arum said that private organizations such as Kengo are useful in moving the development process by helping local communities with problem solving along "a whole continuum of issues." For example, it interviewed people in various areas of the country, found out what they are actually eating, and suggested how they could grow those crops instead of using the land for cattle and other export commodities. It showed people how planting trees could bring them benefits, such as fruits to eat or wood for cash-producing carvings. "People won't plant trees unless they benefit from doing so," he said. Kengo also helps farmers grow crops with low input of fertilizers. "We are getting low prices for our coffee. Should we continue to try to get you in the rich countries to give us a better price or should we stop growing coffee and grow things we need to sustain ourselves?"[4]

. . .

The Okavango River, flowing southward from Namibia, peters out in the hot sands of Botswana's Kalahari Desert. Before it does, however, its delta creates a lush wetland ecological system inhabited by some of the richest and most diverse populations of wildlife remaining in Africa. Human civilization has not yet overwhelmed the delta. It is far from major population centers, and its glass-clear waters, pure enough to drink, can be negotiated only by shallow dugout canoes poled by local villagers familiar with the labyrinth of channels and islands. Although it is difficult and expensive to reach, the Okavango has a thriving tourist trade because of its still abundant animal, bird, and plant life and the exotic beauty of its landscape.

As a dugout turns a bend in the river, a bull elephant rises from a deep pool, its wrinkled skin a glistening black from the water. A profusion of birds create an iridescent rainbow as they flit through the air. Crocodiles slither through the papyrus reeds along the bank. On the grassy islands small herds of gazelles, giraffe, wildebeest, zebra, and an occasional stately kudu feed on the high grass. A band of baboons swing overhead through the sausage trees, moving slowly but discreetly away from a small group of human hikers. The air is fragrant, and on a warm, sunny midday not a sound is to be heard except the wind.

Unlike the plains of the Masai Mara, the Okavango Delta shows little physical impact from tourist traffic. The tourist industry and the local population, with the concurrence of the Botswanan government, has opted for a strategy of low-density, high-priced tourism. Safari camps are relatively small and scattered, and although comfortable, most of them do not provide luxurious amenities such as swimming pools and cocktail lounges that are to be found at many camps elsewhere in Africa. Access to the camps is chiefly by small aircraft, an expensive way to travel. While tourism brings in a substantial amount of money, it makes relatively little intrusion on the wildlife and virtually none on the landscape. A substantial percentage of the money stays in Botswana, according to government officials in Gaborone, the capital city. Were it not for the threat posed by poachers, the tourist industry obviously could continue to provide a modicum of prosperity to this thinly settled area for the foreseeable future.

Several years ago, the Botswanan government drew up plans to draw large volumes of water from the Okavango to expand the diamond mining and cattle industries that have made the country relatively prosperous by African standards. But the tourist industry, joined by area residents and Botswanan and native international conservation groups, objected, contending that for the long-run prosperity of the country, as well as to preserve an irreplaceable national resource, the water should be reserved for the delta and its wildlife. Although the diamond and cattle industries are influential political powers within Botswana, the water has not been diverted—at least not yet.

In 1986 the Botswanan government began drafting a national conservation strategy that would look for ways to enhance economic growth while preserving the natural resource base. In drawing up its plan the government consulted heavily with local communities through what were essentially town meetings held all over the country. While the plan was influenced by the powerful diamond and cattle interests, it also reflects what the communities perceive as the kind of economic development that will meet their interests over time. According to Kagiso P. Keatimilwe, spokesman for the National Conservation Strategy Coordinating Agency, the national plan is based on the premise that "development and conservation cannot be separated." Much of the plan is based on the sustainable use of indigenous plant and wildlife resources as well as on conservation and allocation of the scarce water supply.[5]

While Botswana's national plan was developed by the government, it is carried out largely by local communities with the assistance of NGOs. For example, Thusano Lefatsheng, a nonprofit rural development organization with headquarters in Gaborone, provides agricultural research and extension and marketing services to communal groups around the country. One of its current projects is to help the seminomadic hunter-gatherers of the Kalahari develop an agrarian life-style as development encroaches on their communal lands and their food sources dwindle. Goagakwe Phorano, general manager of the organization, said that it is teaching the Kalahari dwellers how to grow plants indigenous to their land, such as the Kalahari devil's claw, from which a medicine widely used against arthritis in many parts of the world is extracted on a commercial basis. After doing research on how to grow the wild plant as a crop, the group formed cooperative producer groups and is marketing the medicine in places as far away as the United States and South Korea and producing a reliable cash income for people who had almost literally nothing.[6]

. . .

In neighboring Zimbabwe the showpiece of sustainable development is the Communal Areas Management Program for Indigenous Resources, known nationally and internationally as the CAMPFIRE program. Under the program, management and de facto ownership of the wildlife that inhabits the communal lands stretching south and east of the broad and beautiful Zambesi River is being turned over to the communal peoples. The chief goal of the program is the economic benefit of the people. But the project was also designed for conservation of the area's highly threatened wildlife. The result, according to Robert Monro of ZimTrust, a nonprofit economic development group based in Harare, is a textbook example of how environmental and development goals can and must mesh.[7]

Wildlife in the region, as in much of Africa, has been under desperate pressure. The eradication of the tsetse fly in the Zambesi valley had permitted the encroachment of cattle ranches in an area from which they had previously been excluded by disease. Poaching was intense, particularly of the country's threatened rhino population but also of elephants and much other wildlife. Zimbabwe's underfunded and understaffed Department of National Parks and Wild Life Management has been engaged in a fierce war with poachers, with a substantial number of casualties on both sides.[8]

The premise underlying the CAMPFIRE program is that turning over the management and benefits of wildlife to the communal people gives them a strong incentive for protecting and preserving that wildlife, an incentive that was lacking when government policy simply tried to regulate hunting and when cattle ranching was one of the few economic alternatives. The Park Service conducts regular wildlife surveys, provides technical and marketing assistance, and offers advice on optimum harvesting levels and quotas. Zimbabwean governmental agencies as well as national and international organizations are active in providing technical assistance and general support to build the capacity of the rural communities to make informed decisions. Considerable funding has been provided from a wide variety of sources, including governmental development assistance agencies in North America and Europe. But the actual decisions on wildlife use: which and how many animals to kill for food, to sell for hides and meat and other products, or to offer for sports hunting, are made by the communal councils—by the local people themselves.

Simon Metcalfe, a Zimbabwean scholar who has studied the CAMPFIRE program closely, said that it holds the view that "community awareness is based on proprietorship and once the structural conditions for participation are correct the households, villages and wards will become motivated by the prospect of increased income security." He added that "CAMPFIRE attempts to provide a means to harmonize the needs of rural people with those of ecosystems."[9]

Rob Monro put it a bit differently. "In the developed world," he said, "there are two mental pictures of Africa. One of the images is of drought, starvation—the Somali image. On the other hand, there are equally clear images of this wonderful Garden of Eden with elephants, giraffes, and antelope. And, in fact, it is the same Africa. But people outside Africa are not able to put the two together in their minds. What CAMPFIRE has been doing is putting the two together. CAMPFIRE is the melting pot where the two come together." One lesson from this, he said, is that conservationists and others in the North should not seek to ban hunting and trade in wildlife products because only by making wildlife economically valuable to Africans will it be saved.[10]

. . .

These kinds of bottom-up development projects, depending on local planning, carried out by communal institutions, making use of the formerly neglected expertise of women and the wisdom of elders of the community are sprouting throughout the continent, said James Kamara, program officer for the African

Ministerial Program on the Environment. Frequently they rely on native crops or on indigenous wildlife and are planned to fit local ecological and economic conditions. These projects are a first step toward relieving the food scarcity endemic to much of Africa and are also introducing environmentalism to Africans at the grass roots.[11]

Such projects probably meet most of the many varying definitions of sustainable development. But given the overwhelming problems facing most African nations, including rapid population growth, unemployment, urbanization, poverty, hunger, disease, inadequate infrastructure, illiteracy, and ethnic strife, not to mention depleted national treasuries and high external debt, such small-scale development by itself, however sustainable, is inadequate, Kamara and other African experts agree. E. T. Mundangepfupfu, Zimbabwe's minister of the environment, recounted how his government had called a national conference after Rio to draw up its own Agenda 21. The participants concluded that while tourism and wildlife utilization were promising sources of economic growth, they could not by themselves even begin to meet Zimbabwe's needs. Instead, what would be required was an economy restructured to promote industrial development. To do so, however, would require the kind of additional financial and technical assistance from the North called for in the Agenda 21 issued at the Earth Summit, Mundangepfupfu said. "So far," he added, "we have only promises."[12]

. . .

Experiments in sustainable development at the community level, carried out by local residents and supported by NGOs, are springing up on every continent, in virtually every country. Chiefly they are taking root in the poorer countries of Africa, Latin America, and Asia, but they are also found in North America and Europe. They are widely diverse in their structure, goals, and methods. All of them share the traits of being run by local people for local people.

. . .

Eastward from the Ecuadoran coastal city of Esmeraldas, the rain forest has been cleared to make way for cattle ranches, leaving behind a vast expanse of bare brown soil. For many miles around the Communa Rio Santiago Cayapas, however, the thick green canopy remains intact. It is from the trees and the life that lives beneath and within them that the villagers of the *communa* obtain their food, their shelter, and most of their clothing and other worldly needs.

The villagers are among the poorest in Latin America. The average family income is less than the equivalent of $80 a month. Infant mortality is high, so is the incidence of childhood diseases and other illnesses, such as malaria, cholera, and river blindness. In one sense, however, the villagers live amid great wealth. Biologists estimate that the surrounding ecosystem contains between eight thousand and nine thousand different species of plant, more than in any ecological system of comparable size in the world. The women, men, and children of the *communa* harvest the fruit, nuts, medicinal plants, and game and cut the trees sparingly for their houses, their canoes, and their cooking fires.

In time the forest that provides a living for the people of the *communa* could be lost to the advancing chain saws to provide lumber for the global market and profits to wealthy entrepreneurs who invest in cattle ranches. But recently, the villagers have been given an alternative—a way to make the forest produce a cash income and that makes it worthwhile to save the trees.

In 1990, Conservation International, a Washington, D.C.–based environment and development organization, offered the *communa* a plan to enable their villages to survive and perhaps even prosper while leaving the trees standing. The plan was based on the nut of the tagua tree, which has the hardness and white sheen of ivory. Earlier in the century the tagua nut was widely used as an inexpensive alternative to ivory buttons for shirts and dresses, until it was replaced by cheap plastic. With the international ban on the ivory trade in the late 1980s, the tagua became an attractive substitute for high-quality buttons. Working with local leaders and a local rural development foundation, Conservation International helped organize collectives to harvest, shell, and transport the nuts. It also helped organize local button-making factories and lined up clothing manufacturers in Europe and North America to buy the buttons.

By 1993 more than one thousand people were employed by the project. In three years the *communa* sold more than 850 tons of tagua nuts and generated revenues of more than $1.5 million. Profits were shared by the harvesters, the carriers, the button makers, and the local foundation. Conservation International is now conducting research on other forest products that can be converted into cash without harming the trees and the diverse life they support.[13]

. . .

Siberia's Lake Baikal is the oldest, deepest lake in the world, containing one-fifth the planet's supply of fresh water. Its clear waters and fertile shores are habitat for an extraordinary range of wildlife, including freshwater seals and walrus and fish found nowhere else in the world.

In recent decades the purity of the lake's waters and the productivity of its ecological systems have been threatened by pollution from tanneries and other factories on its littoral and from chemical fertilizer and pesticide runoff from surrounding agricultural projects. Conditions grew so bad that inhabitants of the region began to stage public protests, even when a repressive Communist regime was still in power. Among the protesters were the Buryats, a pastoral people living in the Olkhon region on the western side of the lake. The Buryats, who are a shamanistic, nature-worshiping culture and hold Lake Baikal to be a sacred place, were deeply offended by the desecration of its waters.[14]

Pollution of the lake has slowed since the collapse of the Communist system. Now the Buryat are seeking to develop an economy that will sustain them without harming their environment. Although traditionally a nomadic people, under the Communist regime they had been settled permanently in villages. Much of the land around the villages had become eroded and unproductive because of overgrazing. They grew virtually no fresh fruits and vegetables, depending on processed foods shipped in from distant parts of the Soviet Union to supplement the meat from the sheep they raise.

With a financial boost from the U.S. Agency for International Development (AID), residents of the region recently organized the Center for Sustainable Agriculture. A U.S. NGO, the E. F. Schumacher Society, which is based in Great Barrington, Massachusetts, is helping the Buryats find local solutions to their development needs. It is providing low-key technical assistance to help the local people grow fresh produce and also—using techniques practiced by Amish farmers in the United States, who use no electricity—how to preserve food at home. With another small grant from U.S. AID, the people of the Olkhon district are planning to set up a small cooperative cannery. Susan Witt, executive director of the Schumacher Society, noted the Buryat shepherds had sold their sheep on the hoof, allowing other areas to capture the value added from profits made from processed products. The society is helping the local residents to develop cottage industries that will produce such things as clothes, rugs, and cloth and help them enjoy a greater value for their labor.

The centerpiece of the sustainable development program, Ms. Witt noted, is the introduction of a new land tenure system under which all the land will be held by a "democratically controlled land trust" that will lease land to the Buryat shepherds and farmers. The trust will also provide for private ownership of houses and improvements on the land, such as fences. The land tenure project is part of a much bigger land plan drawn up for the entire Lake Baikal drainage area by George D. Davis, president of Ecologically Sustainable Development, who helped design New York State's strategy for protecting the Adirondack Park.[15] In the Olkhon district, which embraces 6,000 square miles and 9,500 people, the land tenure plan will encourage private investments in businesses and housing while assuring that the essential characteristics of the unique Baikal ecosystem will be preserved, Witt explained. As an interesting aside she commented that "this project would not be possible without E-mail." Because of the difficulty in communicating with that remote district of Siberia, U.S. technical advisers find that the only way they can be in day-to-day touch with the Buryat managers of the project when they are not actually there is via Internet. Here is a lesson in appropriate technology![16]

Some observers see the Olkhon community trust as a potentially path-breaking model for a "the peaceful decentralization of a failed and crumbling nation-state."[17]

. . .

Colette Machado, a large, outspoken woman who lives on the Hawaiian island of Molokai, is a fierce advocate of self-determination for the remaining native Hawaiians. She is also resolved to preserve the unspoiled rural character of her island and to keep it from becoming, as other of the islands have, an overdeveloped, urbanized, expensive tourist paradise from which much of the native flora and fauna, as well as the native way of life, has been extirpated. Both goals, she decided several years ago, require that the native people of Molokai develop their own economic support system. A schoolteacher by profession, she has organized a cooperative aimed at encouraging the development of small-scale commercial enterprises that rely on local resources. The name of her organization is Ke Kua'aina Hanauna Hou, which roughly translates as "the new generation of traditional Hawaiians."

The centerpiece of the project is the twenty-eight-acre Puko'o beach front and fishpond on the eastern end of the island. The artificial pond, whose blue waters are alive with fish and sea turtles, is used to cultivate an edible seaweed, called lipu, which Ke Kua'aina learned how to grow and harvest with technical help from Arizona University. The seaweed is sold fresh in Honolulu and other parts of the state and is also processed as soup, pickles, and a kind of sauerkraut called kim chee. The cooperative has its own retail operation for selling these products as well as other products processed from locally grown crops, including jams, jellies, relishes, dried fruit, and sweet potato chips. The project, Machado said, actively encourages and assists Molokai residents to cultivate traditional crops such as taro and breadfruit and helps them to market their surpluses. It is also encouraging the restoration of the artificial fish ponds built into the ocean by Hawaiians before the Europeans arrived and making them viable sources of food and income. Today the ponds, many surrounded with mangrove trees, often have breached dikes in need of repair. Machado and other activists in the drive for self-rule for native Hawaiians made access to the ponds, many of which are centuries old, one of the demands of their self-rule campaign. She also takes pride that her group helped block a large condominium project planned by developers for the South shore of the rural island.[18]

"I want people to start developing their own revenue doing traditional activities like fishing and farming so they don't have to sell their land or move into tourism," Machado explained. "We've empowered the community into developing a policy with a strong preservation message."[19]

. . .

With its burned-out buildings, its garbage-strewn vacant lots, its high unemployment and crime rate, its drugs and despair, New York City's South Bronx has long been regarded as a prototype of urban failure in the United States. As Hispanic- and African-Americans moved into the area in the decades after World War II, middle- and working-class whites fled, and with them went virtually all of what had been a reasonably thriving manufacturing economy.

Kelly Street, a small avenue curved in the shape of a banana, was just another byway in a badly decaying neighborhood. But during the mid-1970s residents organized to block the planned demotion of three abandoned buildings on the street. Without tools or money or title to the property they began to rehabilitate the buildings to save them from the wrecker's ball. Against odds, they succeeded, and their success inspired them to form the Banana Kelly Community Improvement Association, a grass-roots community development corporation with a goal of revitalizing the South Bronx by creating new economic opportunities. But, wrote Yolanda Rivera, a single mother of two children who is chairperson of the association, "One day, 20 years later, we found no jobs, no industry, terribly contaminated land, high mortality rates, learning disabled children, garbage everywhere and waste treatment facilities that violated the environment and quality of life standards we were working so hard to raise."[20]

Looking for a business opportunity that would protect the environment and be useful to society at large, as well as provide jobs and invigorate the community, Banana Kelly linked up with the Natural Resources Defense Council (NRDC), a respected national environmental group with headquarters in Manhattan that had previously concentrated most of its efforts on research, litigation, and lobbying. They decided to join in a partnership to create the Bronx Community Paper Company, which would take waste office papers from New York's business community, de-ink the paper, and turn it into pulp, which would be recycled with virgin pulp as high-quality office paper. A mill is planned for a nineteen-acre site in the Harlem Rail Yard that is owned by the city Department of Transportation and is leased by a consortium of private companies that wants to transform it into an industrial park. It is estimated that the plant will employ some two hundred people and create support jobs such as day care and medical facilities and other social services for an additional one hundred.[21]

The project attracted the attention of the Modo Paper Company of Stockholm, which has contributed $300,000, and S. D. Warren, a subsidiary of Scott Paper Company, which has given $250,000. "This is not a social project for us—we need the pulp," explained James D. Black, vice president of S. D. Warren.[22] While changes in management at both companies have complicated negotiations, the project was still moving forward in 1995. The NRDC has provided $770,000, and the state of New York has given $1.1 million.[23] In all, some $260 million will be needed to build the plant and allow it to commence operations.

Allen Hershkowitz, a senior scientist with the NRDC and, with Rivera, the co-chair of the Bronx paper company, asserted that the Banana Kelly project represented "the key to the environmentalism of the future. If a sustainable economy didn't exist, then it will have to be built. We have to figure out how to make the economical growth of the future environmentally sound. We have to outcompete projects that are environmentally bad. Everyone is an environmentalist if you can make it profitable for them." A working definition of sustainable development, Hershkowitz said, is "industrial ecology plus social justice."[24]

. . .

Although they are in very different countries and cultures and physical and economic settings, all of these development projects have certain common characteristics. They are relatively modest in scale, they tend to be labor intensive, and with the exception of the urban Banana Kelly project, they do not require large inputs of capital, expensive equipment, or infrastructure. In many ways they are the antithesis of the big sectoral development projects financed by the multilateral development banks or bilateral aid programs, such as the Volta Dam in Africa, which flooded thousands of acres of land, displaced thousands of local residents, and caused disease and cultural dislocation, all in order to provide electric power for a foreign-owned aluminum-producing factory. All of the projects require not only the consent but the active participation of the local communities they will affect.

But these small projects usually do require some outside assistance in the form of seed capital, technical assistance and small-scale technology for tools, and especially information about methods and techniques. They generally make use of local plants and wildlife and other indigenous resources. (In New York the indigenous resource is the huge volume of waste paper generated by business.) These resources are renewable but only if they are husbanded and used wisely and carefully. Participants in the program, therefore, must be sensitive to their surrounding environment and aware that they must preserve that environment to prosper. They become, as conservationalists David Western and Michael Wright note, beneficiaries as well as custodians of conservation.[25]

But members of the community also must have a permanent personal or communal stake in the project. These examples suggest that not only must people receive the benefits of the economic activity and a role in planning and management of the development, they must also have a vested interest in the form of control of the resources, land tenure, or part ownership of the project (shares in the corporation). With such an incentive, participants—once they are beyond the struggle for mere survival—are likely take a long-term view. Because it is theirs, they are concerned about the future of the local ecosystem and its resources, not just about quick returns and maximizing profits.

The examples cited in this chapter and similar projects proliferating in every part of the world make clear that sustainable development is not simply a matter of applied economics or science and technology but, in a phrase used by Stephen Viederman, president of the Jesse Smith Noyes Foundation, a "social construct." It is a structure that links people's basic economic needs and cultural patterns with the long-term welfare of their physical surroundings through a democratic, participatory process.

Looking at development from this perspective, Viederman has reached a substantially more detailed definition of sustainability than the one offered by the Brundtland report. His definition:

> Sustainability is a participatory process that creates and pursues a vision of community that makes prudent use of all its resources—natural, human, human-created, social, cultural, scientific, etc. Sustainability seeks to ensure, to the highest degree possible, that present generations attain a high degree of economic security and can realize democracy and popular participation in control of their communities, while maintaining the integrity of the ecological systems upon which all life and all production depends, while assuming responsibility to future generations with the wherewithall for their vision, hoping they have the wisdom and intelligence to use what is provided in an appropriate manner.[26]

It must be recognized, however, that even small-scale, community-based sustainable development requires certain preconditions and is not always possible. One crucial issue is how to disengage a community—or a nation for that matter—from an economy of survival so that it is able to build a sustainable economy. If a family is starving, it must eat the seed corn. Fabio Feldmann, who is the first environmentalist to be elected to Brazil's House of Representatives, found that "the basic contradiction to overcome is the one between the assumed sacrifice of nature in the name of hunger, and the need for environmental planning aimed at avoiding, in the future, the occurrence of an irreparable nutritional deficit." Feldmann also warned that "one must avoid unwarranted expectations of the results of local action, underestimating the importance of investing in national and international institutional building." He noted that democratically controlled community-based development is difficult or impossible when the state does not countenance democratic institutions or "winks" at the predatory use of local resources by favored economic or political interests.[27]

Community-based economic development cannot take place, of course, without a viable community. In many areas such communities do not exist or are in the process of dissolution: witness the disintegration of social cohesion in Rwanda and in Bosnia, among other troubled areas of the world. Moreover, experience in recent decades has shown that success in economic development, particularly development on the Western model, can lead to the weakening of traditional communal values. Certainly communal ties in many industrialized countries, especially the United States, have weakened considerably since World War II.

Efforts at local sustainable development can be overwhelmed by high population growth, which makes sustainability an ever receding target. Also, too many people can disrupt communities internally through crime or erode community cohesion by forcing migration. Small communities, too, can be subject to corruption, rampant commercialism, nepotism, and other cultural failings that undermine efforts to improve their quality of life and preserve their land and resources. AIDS or drugs or other massive public health problems can sap a community's strength to concentrate its resources and collective will on long-term economic activity. Local development may be undermined by the lack of adequate institutions, capital, information, education, scientific and technical skills, or other social resources. And as Viederman points out, there is no "politics of sustainable development" to support community efforts, only a politics of special interests.[28] While there is an increasingly persuasive economics of sustainability, such economics is rarely the basis of policy decisions at any level—at least not yet.

Even if community-based sustainable development projects were to spread across the globe like a prairie fire, they could not, by themselves, end the planet-wide twinned economic and environment crises. Countries such as India and China, with millions to feed, clothe, house, and educate will need large-scale economic growth meet those needs. Not all of the one-fifth of the world's population that is in absolute poverty can be extracted from that condition by small, ecologically sound economic projects—certainly not the swelling millions flocking to already crowded and impoverished cities.

The modest economic goals represented by the projects described in this chapter would not satisfy the aspirations of all members of the human community. The native Hawaiians of Molokai may be seeking a return to a simpler, older life-style. Islamic fundamentalists may reject Western materialism. But it is probably fair to say that millions, perhaps billions of other people in the world see the affluent consumer-oriented economies of the rich industrialized countries as their model. Beyond economic security they covet a comfortable place to live, good food and clothes, a television set, a telephone, an automobile, leisure, and travel. Many nation-states crave the kind of political and military power that accompanies that level of economic development.

As discussed in earlier chapters, many experts insist that the planet simply could not sustain 5 billion people living that kind of life-style—not to mention the 10 billion people the planet will have to support in the next century. Indeed, there is a real question as to whether the current levels of consumption and pollution generated today by a handful of affluent industrialized nations is sustainable. In a number of areas, including the production of greenhouse and ozone-depleting gases, these economies have already overshot maximum levels that are compatible with a healthy planet. Local community action on low-lying island nations will not hold back the rising waters that may inundate them as ocean levels rise with global temperature. The technologies, production methods, marketing practices, and wastes of the great transnational companies could easily overwhelm progress toward sustainability in the developing countries if the management of those companies is insensitive to local and national needs.

Some observers, especially biologists, contend that sustainable economic development for humans, even small-scale development at the local level, is incompatible over the long run with the sustainability of other life on the planet. John Robinson of the New York Zoological Society wrote: "Sustainable use is very appropriate in certain circumstances, but it is not appropriate in all. It will almost always lower biological diversity whether one considers individual species or entire biological communities, and if sustainable use is our only goal, our world will be the poorer for it."[29]

Local, small-scale, bottom-up community development clearly represents progress down the road toward meeting the environment-development crisis. It can alleviate poverty and despair. It gives participants a sense of possibility, a sense that they are working for a better future for themselves and their children. Rising affluence, even on a modest scale, has been shown to dampen population growth and to ease pressures on local resources. Such development can lessen the economic need for energy- and resource-intense agriculture and industry.

But it certainly cannot eliminate that need for the foreseeable future. As the authors of *Saving the Planet* admonished, we are still failing on a global scale "to alter the basic patterns of human activity that cause our environmental deterioration."[30] We are still practicing the same kind of economics, industry, and commerce; employing the same kind of science and technologies; engaging in the same kind of politics; looking at our security from the same narrow nationalistic, even tribal, perspectives. With signs of slight modification here and there, individual humans are still engaged in the same kind of reproductive and consumption patterns that have brought us into the global pickle barrel.

In the meantime the planet continues to be afflicted by the intertwined evils of pollution, depletion of its storehouse of natural resources, damage to the systems that support life, and economic inequity, poverty, and human suffering. At the Earth Summit it was agreed that we are running out of time to do away with those evils, that they no longer can be left to the next generation. As the actor on the public television science program says every week, "The future is where we will spend the rest of our lives."

Chapter 14

June 2042

The future enters us, in order to transform itself in us,
long before it happens.
—Rainer Maria Rilke, *Letters to a Young Poet*

Adam, my first grandchild, was born in the spring of 1991, a year before the Earth Summit. In June 2042, the fiftieth anniversary of the summit, he will be only fifty-one years old. His sister Sophia and his twin cousins Edward and Alexander, born after the summit, will be in their forties. I will be long gone by then, but these lovely little humans, who are here with me today and for whom I care deeply, will be in the prime of life. So I think a lot and worry a great deal about a future that I will not see.

Some people profess confidence that the future will take care of itself. For example, Julian L. Simon, a professor of business administration at the University of Maryland and a vocal skeptic about environmental threats, contends that "almost every trend that affects human welfare points in a positive direction, as long as we consider a reasonably long period of time and hence grasp the overall trend."[1] He argues that history demonstrates humans have steadily improved the condition of life on earth, and therefore, the more humans there are, the better off we will be.

I would like to be able to take such a Panglossian view of the future. But that perspective is very much like the French approach to defending themselves from Germany after World War I. Assuming on the basis of recent history that the next war would also be decided by the strength and ingenuity of fortifications and entrenchments along a fixed front, the French built the Maginot Line. It was, of course, quickly overwhelmed by the new tactics of Nazi Germany—the mobile, mechanized, land and air blitzkreig. Within weeks, France was a defeated, occupied country.

One of the lessons of history is that it is filled with discontinuities, that the past is not necessarily prelude. Many thoughtful observers, looking at the environmental and economic trends of this century believe we are in or approaching just such a discontinuity. The physicist M. L. Budyko wrote nearly twenty years ago: "The development of civilization, an exceptionally rapid event on the time scale of geological history, has radically changed the outlook for the future existence of the biosphere."[2] Eric Ashby, the first chairman of Britain's Royal Commission on Environmental Pollution, cautioned that "what we are experiencing is not a crisis: it is a climacteric. For the rest of man's history on earth, so far as one can foretell, he will have to live with problems of population, of resources, of pollution."[3]

Simon does have one point. Humans do have the ingenuity and capacity to address all of the many problems described in this book and to create something resembling an ecologically sustainable global economy and a peaceful global society unified in their commitment to preserving the planet and sharing its fruits equitably. Millennial *angst* about the dire fate awaiting the planet and its inhabitants is at least premature.

But the notion that human adaptability, along with science and technology and a somehow infallible economic system, will automatically take care of our future strikes me as akin to the religious belief that faith alone will assure salvation. Perhaps it is true, but the stakes are so high that it is probably prudent to back our faith with good works. A change of course is necessary. It will not just happen, as Simon and his libertarian ilk appear to think. The invisible hand of free-market capitalism, unconnected to a guiding intelligence, is not by itself strong enough to push the human community in the right direction. Human numbers and ingenuity will not save us if a majority those humans are submerged in a sea of poverty, hunger, disease, and ignorance and if the rest pursue lives of self-obsessed consumption and individualism. A transition to an environmentally and economically sustainable society will require just that—a transition. We humans, along with our governments and other institutions, will have to act, to change, to overcome inertia in order to make that transition. To entrust our future to luck, to fate, or to blind faith in a manifestly flawed economic system would be inexcusable folly.

By now we know what needs to be done. The agenda for the twenty-first century agreed on at the Earth Summit provided some eight hundred pages of specific instructions. Libraries are filling with other thoughtful prescriptions for saving the planet. These may not be complete or precise answers. There is still a great deal to learn and to think about. But if we are still heading in the wrong direction, it is not because we lack road maps.

We know that we have to stabilize and probably reduce population levels to make our goals of a healthy, prosperous, comfortable, diverse, more pleasant, and still beautiful planet achievable. We know that we must end the assault on the Earth's life-support systems with our wastes and pollution, poisons and radiation, that we must use our air, water, soil, and other resources wisely and carefully and husband our genetic legacy. We are aware that it is in our own interest, as well as our duty, to share the wealth of the planet more equitably, not only with its current co-tenants but with future generations as well. We will have to develop and deploy technologies that cooperate with nature rather than deplete and destroy it. Our economies must develop within a budget set by what the natural world can provide without becoming harmed or exhausted. To safeguard both our habitat and our livelihood we must cooperate as individuals and as nations.

In the world as it has become, as Sir Crispin Tickell, elaborating on John Donne, observed: "No man is an island, no island is an island, no continent is an island."[4] The world leaders in Rio agreed that collective security in the future will hinge to a substantial degree on the ability of nations and peoples to act collectively to achieve a sustainable society for all.

Most of us have also accepted, I think, that the health of the environment and the health of the economy are inextricably engaged, that the natural world and the world that humans create to live within it should form a seamless continuum. Less widely recognized perhaps is the fact that our economic and ecological welfare depends also on the political health of the planet. My measure of political health is the degree to which individual citizens participate in the decisions that rule their lives and the abundance of democratic institutions, including civic institutions such as environmental groups and local community development associations.

One of the things that struck me again and again as I did the reporting for this book was that effective action to promote economic growth or environmental protection, or both, almost always takes place within a democratic context. Democratic nations are those that sustain strong economies and protect their environments over the long run. Even in the absence of a democratic national government, development and resource preservation can take place at the local level when there is citizen participation in the planning, decision making, operation, and allocation of rewards. In the developing world it appears that those countries that have made the greatest progress toward a democratic society also have achieved the most economic progress and are farther along the road to environmental stewardship. Conversely, those countries that have lagged in establishing democratic institutions and liberating the energies of their citizens are the ones experiencing economic hardship and dangerous, deteriorating environments.

There are other necessary dimensions. Even functioning democratic institutions will not be able to withstand the multiplying pressures created by human activity unless they become wiser and more just and equitable. Democracy without equity is highly vulnerable. And finally, good citizenship is grounded in love of place, in care for the natural habitat that sustains us.

Exceptions can be found. China has managed to achieve an economic boom while repressing democracy and human rights. But its environment is badly blighted. And its economic prosperity probably is either temporary or the prelude to greater democratization. The events of the latter part of this century strongly suggest that the absence of democracy is not sustainable on either an economic or an ecological basis.

. . .

That we know what must be done does not mean, however, that we will be able to do it. The Earth Summit process and its aftermath amply demonstrate that the path to an ecologically secure and economically prosperous and just future is filled with formidable, perhaps insurmountable, obstacles and bottlenecks. These obstructions are everywhere, from our reluctance to create a workable system of global governance to the seeming inability of many, if not most, individual humans, whether out of ignorance or indifference or selfishness, to raise our line of sight above the near horizon. The prevailing global approach to the environment-development dilemma still is well described by the comment of Peter Kalm, a Swedish naturalist traveling in the United States in the mid-eighteenth century, who, appalled by the Americans' waste and destruction of resources, wrote: "Their eyes are fixed upon the present gain and they are blind to the future."[5]

Judging by what has and has not happened since Brazil, the will and ability of the human community to move forward toward the kind of future envisioned in Rio also are burdened by a narrow view of what must be done. Some see the only problem as too many people competing for too few resources, as destructive technologies, or as incapable governments, institutions, and political systems. Others argue that it is excessive consumption by the rich, or the economic power concentrated in the great corporations, or the destruction of the seed corn by the desperate poor, or misguided nationalism that insists on sovereign rights at the expense of cooperation to protect the global commons. Still others define the problem as individual ignorance, indifference, and selfishness or perhaps the increasing alienation of humans from the natural world.

One of the inescapable lessons of the Rio process and its aftermath, however, is that all of these obstacles must be overcome to move toward a sustainable future. And more. To a growing number of economists it is becoming evident that our economics must reflect more realistically what truly constitutes wealth and progress. There is a budding realization even within scientific circles that what is presented as objective science may reflect not truth but the agenda of those who pay for it or the values of those who conduct it. Such science does not always serve the broad public good. Relations among nations are still dominated by the struggle for competitive advantage, for the accretion of power, rather than by a search for ways to cooperate to improve the lot of humanity and to safeguard humanity's habitat. The voices of nongovernmental organizations (NGOs) and of individual citizens are still too little heeded in the councils of national governments and international bodies. The causes of justice and equity within and among nations are too frequently sacrificed on the altars of money and power. And the highest obstacle of all may be the tragically widespread ignorance of or indifference to the jeopardy in which we are placing our posterity.

Clearly, therefore, much will have to change if Adam and his generation are to live reasonably secure and comfortable lives on a healthy, prosperous, and peaceful planet. It will not necessarily be easier for them, even if we started today to do all those things we know we should do. Unless there is a dramatic shift from current demographic projections, the world will be more crowded and could be hungrier by 2042. According to the most credible current scientific consensus, the world will be substantially warmer, with inundated coastal areas, more extreme weather conditions, and unpredictable conditions for growing food. There will still be chlorine in the stratosphere, depleting the ozone that shields us from harmful ultraviolet radiation. Our grandchildren and great-grandchildren will be at least somewhat poorer in their storehouse of genetic materials as human activity continues to eliminate the habitat of nonhuman life. They will not be able to match the material consumption of their grandparents, or at least it will have to be a different kind of consumption. Adam's cohorts in what are now the developing countries will not have achieved the kind of material affluence indulged in by people in the rich countries today, although, it can be hoped, they will have emerged from deep poverty.

We cannot, of course, accurately foretell what the world of fifty years hence will be like. But if governments, institutions, and individuals fail to act, fail to do what we now know must be done, Adam's and subsequent generations may be forced to live hard, even desperate lives. They would inherit a hotter, drier, poorer, hungrier, more polluted, diseased, and biologically impoverished world. They may be forced to live in a smelly, garbage-strewn, paved-over, poisonous landscape, where trees and open space are rare and from which wildlife has been extirpated—except for vermin. It could be a world of intensifying competition among nations and individuals, not for riches but for the means of basic survival. It could be a world with social and political relationships stressed to the breaking point. Democratic institutions could decay, and individual liberty could be ground down by the harshness of life.

Writing in the *Atlantic Monthly,* journalist Robert Kaplan contended that in parts of today's West Africa can be seen the pattern for the coming "worldwide demographic, environmental and societal stress, in which criminal anarchy emerges as the real 'strategic' danger. Disease, overpopulation, crime, scarcity of resources, refugee migrations, the increasing erosion of nation-states and international borders, and the empowerment of private armies, security firms, and international drug cartels" are what is in store for the human community in the early decades of the twenty-first century.[6]

Some forecasters warn that without a change of course, even rich, powerful countries such as the United States will be unable to insulate their citizens from the physical and economic decline of the planet. In their book *How Many Americans?* demographer Leon F. Bouvier and Lindsey Grant, a former deputy assistant secretary of state for environment and population, warn that

> if the nation insists on staying the present course, we will be living the simple life in 2050—and it won't be as attractive as the old song suggests. We won't have cars, and without the physical security that cars provide, we won't be venturing out much in that turbulent society at night, except as part of an armed group. We will be paying higher taxes for such security protection as is available, and for getting the poisons out of the air we breathe and the food we eat. . . . We will drink the sewage from neighbors upstream. . . . If we are unlucky and nobody has prepared for the energy transition, we may be cold and huddled beside a single light bulb. . . . We will probably be on a diet closer to that of 1900 than of 1994.[7]

But Adam need not face that kind of future. If wise decisions are made and decisive action taken, if politicians, business leaders, economists, and consumers can escape from what the Brundtland report called "the tyranny of the immediate" and think about and plan and act for his generation, then he will not have to live in the kind of dismal world pictured above. He could enjoy, as his grandparents' generation has, expanding personal prospects, a longer, healthier life, greater freedom and security, and an improving—although very different—standard of living. He may face some problems, such as a warmer planet, that we have not, but overall, his generation ought to be able to continue the upward progress of humankind.

The knowledge and means for such progress are already pretty much in hand as the new century approaches. The economies of North America, western Europe, Japan, and several other Pacific Rim countries are already poised at the edge of the postindustrial era, when information, communication, and robotics will not only continue to ease humanity's labor but could also sharply reduce the need to pollute and poison the environment and squander its resources. In a world where the widest spectrum of information is broadly accessible, there should be no lack of understanding anywhere about the dangers facing the human community and the solutions to those problems.

The flowering of knowledge about the genetic structure of life and how to manipulate it can open the way for dramatic new advances in medicine and the prolongation of human life. Such knowledge, used wisely and very cautiously, along with the continuing development of new technologies, could help end hunger on earth—provided, of course, that expanding population does not make that goal an eternally elusive target. New technologies have already prepared us to enter the solar age and enjoy bountiful, cheap, nonpolluting energy—provided, of course, that the controlling economic powers permit us to end our enslavement to fossil fuels, if only gradually.

The emergence of grass-roots economic development programs, particularly in the Third World, is demonstrating that local self-reliance is a feasible goal for a substantial part of the human community in the postindustrial age. We are able to reduce the need to transport people and goods over long distances to service the economy and therefore are also able to reduce pollution; save land and resources, time and money; and allow local communities to make the economic decisions that most intimately affect their daily lives. Such a transition could also ease the pressures on our crowded and unworkable cities. The explosion of democracy around the world in the last decades of this century not only reflect peoples' demand for the right to make their own political and economic decisions but also helps set the necessary foundation for sustainable development. This trend is strongly reinforced by the proliferation of environmental and other NGOs and their slowly increasing influence in many parts of the world.

There are also rapidly expanding networks of international cooperation in many areas, suggesting that the kind of global collaboration needed to carry out the agenda set in Rio is not beyond reach. As the scholar/diplomat Harlan Cleveland has noted, while "most of the news about international cooperation is its absence . . . if you stand back and look at the whole scene you see all kinds of international systems and arrangements that are working fairly well." Among the areas of such cooperation cited by Cleveland are weather forecasting, eradication of infectious diseases, international civil aviation, allocation of the telecommunications frequency spectrum, agricultural research, and cooperation in outer space.[8] A small but growing number of transnational corporations are committing themselves to environmental stewardship and economic equity wherever in the world they operate.

Finally, the end of great power confrontation—if it is not a temporary cessation of hostilities—could free substantial resources as well as the energies of nations and peoples and bring them to bear on the environment/development conundrum.

So a sustainable planet for our grandchildren is not an unreachable goal. If all went well, if we who are responsible for the earth today do all we know we should do, people living a half century from now would be living longer, healthier, happier lives in a clean environment and with an economy that has done away with poverty and reduced disparities of wealth and consumption. If the population tide can be reversed and if wise technology choices are made, *their* grandchildren may be living in a golden age, where there is plenty of space and food, comfort and leisure, freedom and justice, in a rich, safe, regreening natural world.

. . .

Looking at the world as it is today, however, it is difficult to be optimistic that the human community will be able to do what is necessary to attain such a future. The sheer magnitude and complexity of the required transition seems too overwhelming. A civilizational change is necessary: change in the way the global community governs itself, changes in the way nations regard their obligations and privileges, changes in the prevailing economic system, in the oversight of our science, and the conduct and goals of our diplomacy. The fact that more benign sources of energy and gentler technologies are within reach does not necessarily mean that they will be deployed.

And then comes the really hard part.

It is unlikely that a sustainable future can be constructed unless human beings and their societies undergo profound changes—changes in values, in ideas about equity and justice and about how life must be lived and for what purpose. There will need to be a readjustment of fundamental systems of belief, particularly the deeply ingrained belief in nature as a hostile force that must be conquered, possessed, dominated, and exploited. Anthony Cortese, the head of Second Nature, a group seeking to inculcate literacy about a sustainable future, has warned that we must shift away from the "paradigm" belief that humans should exercise dominion over nature and that resources are free and inexhaustible toward a canon of personal and collective responsibility for the well-being of the planet and of our own posterity.[9] To achieve those goals, we will have to minimize competition and aggression among individuals and nation-states and replace it with cooperation.

What is required, it would thus appear, is nothing less than a new stage of human development, a conscious effort by our species to make an evolutionary leap. As the eponymous ethnologists Lionel Tiger and Robin Fox have pointed out, humans are still aggressive and violent primates. Language and tools, they find, have only enabled Man "to turn his primate power struggles into human politics, his primate ecologies into human economies."[10] But those primate qualities that enabled humans to survive and evolve are now working against us in many ways. Our aggression, combined with our use of tools, is destroying the environment we still and always will need to sustain us. And of course, the means we have contrived for acting out our violence have become insanely powerful and efficient. If we are to move forward with the enormous task of creating a sustainable future for our grandchildren, *Homo sapiens* will have to find some way to bid farewell to its own evolutionary childhood.

But in the waning years of this century and this millennium, we seem in many ways to be retrogressing rather than advancing. The eruption of murderous tribal and ethnic rivalries, of divisive nationalism and religious fanaticism, suggests that we may be retreating toward a darker age of human development. The industrialized countries, rather than meeting the commitments made in Rio and taking the steps necessary to achieve global economic equity, are acting to consolidate and aggrandize their own wealth and power. In several European countries, neo-Fascism and xenophobia are emerging from bloody soil, creating new impediments to international cooperation. Far too many humans continue to produce large numbers of offspring, some as heedlessly as our primate ancestors, seemingly without giving any thought to how they or the world will sustain them.

The United States, where environmentalism first flowered and where the Marshall Plan, the first major effort to promote economic development abroad as an act of enlightened self-interest, was conceived, is setting a particularly painful example. At this juncture of history, the United States is the only nation capable of making the UN an effective center of global governance, the only power that can lead the world community along the difficult road to sustainable development. But the politics that has triumphed in this country in mid-1990s has turned its back on such leadership. Instead, it seeks to turn the country inward to a new isolationism that would slash or abandon development assistance, to an old rhetoric that embraces the discredited and dangerous notion that we are forced to choose between the environment and the economy. It is a politics the promotes the private good over the common good, the right of capital, corporations, and property holders to maximize their gain over the duty of all to protect the commons and to share its fruits more equitably.

The structural reform of capitalism, an inescapable precondition of sustainable development that, as Fernand Braudel observed, would "inevitably be difficult and traumatic, requires a context of abundance or even superabundance. And the present population explosion is likely to do little or nothing to encourage the more equitable distribution of surpluses."[11] Braudel may be even more to the point when he asks, "How can one hope that the dominant groups who combine capital and state power, and who are assured of international support, will agree to play the game and hand it over to someone else?"[12]

If, as is now predominantly the case, those who possess capital and state power continue to seek to increase and consolidate that power rather than share it, then the game is lost. What holds true for governments and great corporations is true as well for individuals, at least for the affluent many in the affluent North. As the Australian philosopher John Passmore rightly noted, "there is little hope for us unless we can moderate our desire to possess."[13]

. . .

Hope we must keep, however, if we are to have sufficient resolution to act on behalf of Adam's generation and our more distant posterity.

There are rational grounds for hope. Human folly is not immutable. Cambridge University historian Ralph Buultjens has observed that, over time, many highly negative human behavioral traits and social practices, including cannibalism, incest, slavery, dueling, and child labor, have either been eliminated or greatly reduced because of social and moral pressures.[14]

Who would have guessed after Hiroshima and Nagasaki that fifty years later not a single additional nuclear device would have been detonated in an act of war? Who would have guessed ten years ago that the nations of the world would enter into treaties to protect the earth's atmosphere and preserve its biological diversity?

The Earth Summit demonstrated that political leaders must now at least acknowledge the need to address the ecological and economic future of the planet. A few even appear even to understand and accept that it is their responsibility to lead the way into that future. Vaclav Havel, the playwright/president of the post-Communist Czech Republic, affirmed: "It is not that we should simply seek new and better ways of managing society, the economy and the world. The point is that we should fundamentally change how we behave. And who but politicians should lead the way? Their changed attitude toward the world, themselves and their responsibility can give rise to truly effective systematic and institutional changes."[15]

While the politics of predatory self-interest is now dominating the United States and some other parts of the world, Lynton Caldwell has suggested that the current "libertarian explosion," with its emphasis on individual rights at the expense of the common good, could be an "interlude between the dying . . . of the old world and the birth of the new."[16]

The post–Cold War peacekeeping efforts of the UN, unsatisfactory though most of them have been, at least made the possibility of global governance seem less chimerical. As UN secretary-general Boutros Boutros-Ghali reported, "There are signs that the system of collective security established in San Francisco nearly 50 years ago is finally beginning to work as conceived and that it is proving able to respond flexibly to new challenges. We are on the way to achieving a workable international system."[17] The way is obviously still very long, but at last we have cause to hope that it is open. In the meantime, the issues of environment and development are already deeply embedded in the international agenda.

There is also hope to be found in the widening influence of ecological economics in business, political, and intellectual circles, despite the dismissive skepticism of some mainstream neoclassical economists. The marriage of ecology and economics looks to be a lasting one. It also appears that a democratizing and increasingly humanized science may be up to the challenge of helping to extricate humans from the ecological pickle barrel we have gotten ourselves into by the destructive and nonsustainable application of science.

It is also at least possible, finally, to envisage the essential evolutionary jump of human intelligence and emotion into a postmodern era emerging from a synthesis of science with a reawakened transcendental regard for the nature. The modern world, built on the edifice of scientific materialism, transformed nature into something remote from and almost extraneous to the human condition. Modernity freed humans from the fear and awe of natural forces and from the limits that nature placed on how we might think and act. It replaced the reverence for natural world and the transcendental links with its creator or its essence with a with a pragmatic, thoroughly irreverent, and unromantic view of the earth as a storehouse of materials to be studied, developed, and put on the market.

But there was a price to be paid. As the historian Franklin L. Baumer noted, it was widely held among at least one school of intellectuals in the twentieth century that the freedom of man in the modern world "doomed him to live like a stranger in an indifferent universe."[18] Aliens on their own planet, humans were free to exploit and abuse nature—and each other.

Over the course of this century, particularly in its last decades, the knowledge provided by science has disabused us of the notion that nature is indifferent to our actions and that we may safely be indifferent to nature. It has become apparent that our supposed freedom from nature is threatening our physical health and safety and our economic security and comfort. This knowledge can point us into a postmodern era in which a restored transcendental sense of unity with nature reinforces rather than conflicts with the application of science and technology to sustaining us and solving our problems. If we love the planet, we will want to use it wisely and gently. Science and technology can enable us to do so while still meeting our material needs.

Those needs would be somewhat redefined in a postmodern society. Maurice Strong called for the adoption of life-styles of "sophisticated modesty," in which the enjoyment of leisure and the cultivation of the arts would replace to some degree the consumption of resources and the accumulation of things as goals of economic activity. The writer Ferenc Maté suggested that most of us could live, as the title of his book stated, *A More Reasonable Life,* by which he means a life of more self-reliance, of attachment to family and community, and with a sense of identity with place—a life lived among trees and gardens, as opposed to a life dedicated to the consumption of ephemera and dependent on complex, remote economic and industrial systems and lived in an impersonal, denatured urban setting.[19]

Of course, many people, particularly in the developing world, live such "reasonable" lives and yet go hungry and ill-housed and clothed, in ill-health, poorly educated, and forced to degrade their local environment in order to survive. But a more reasonable life led by the affluent, overconsuming minority of the rich industrial countries could free sufficient resources to permit those with whom we cohabit the planet to also lead lives of modest reasonableness.

Were we able to do that, we could be on the way to achieving Pax Gaia, the Peace of the Earth. The Gaia Hypothesis, propounded by the British scientist James Lovelock, holds that the earth is a single, living, self-regulating organism. Humans, he wrote some fifteen years ago, have not yet accepted that they are part of that organism and do not yet seek to integrate themselves and their societies into the biosphere. But, he added, "It may be that the destiny of mankind is to become tamed so that the fierce, destructive and greedy forces of tribalism and nationalism are fused into a compulsive urge to belong to the commonwealth of all creatures which constitutes Gaia."[20]

At the moment, however, it appears that humanity may have to sink even more deeply into crisis before it will begin to fulfill such a destiny. At the conclusion of their book *For the Common Good,* Herman Daly and John B. Cobb, Jr. find hope in thinking that "on a hotter planet, with lost deltas and shrunken coastlines, under a more dangerous sun, with less arable land, more people, fewer species of living things, a legacy of poisonous wastes, and much beauty irrevocably lost, there will still be the possibility that our children's children will learn at last to live as a community among communities."[21]

. . .

Perhaps. But it is this generation's duty to do all in our power to spare our grandchildren that test.

Notes

Preface (p. x)

1. Pope Paul VI, *Populorum Progressio—On the Development of Peoples* (Boston: St. Paul Editions,), p. 45.
2. "Pope Says Environmental Misuse Threatens World Stability," *Washington Post*, 6 December 1989.

1. The View from Corcovado (pp. 1–8)

1. Conversation with author, 1991.
2. Quoted in Donald Worster, *Nature's Economy* (Cambridge: Cambridge University Press, 1977), p. 30.
3. Quoted in John Passmore, *Man's Responsibility for Nature* (New York: Charles Scribner's Sons, 1974), p. 34.
4. Sigmund Freud, *The Future of an Illusion* (1927), quoted in Barbara K. Rodes and Rice Odell, *A Dictionary of Environmental Quotations* (New York: Simon & Schuster, 1992), p. 197.
5. Aldo Leopold, *A Sand County Almanac* (New York: Oxford University Press, 1943), p. 216.
6. Lewis Mumford, *Technics and Civilization* (New York: Harcourt, Brace & World, 1962), p. 366.
7. Lynton Keith Caldwell, *International Environmental Policy*, 2d ed. (Durham, N.C.: Duke University Press, 1990), p. 9.
8. Edward Goldsmith, *The Way* (Boston: Shambala Publications, 1993), p. 9.
9. Barbara Ward and René Dubos, *Only One Earth* (New York: W. W. Norton, 1972), p. 12.
10. Worster, preface to *Nature's Economy*, p. xi.
11. Al Gore, *Earth in the Balance* (Boston: Houghton Mifflin, 1992), p. 270.
12. Shridath Ramphal, *Our Country, the Planet* (Washington, D.C.: Island Press, 1992), p. 269.
13. "World Scientists' Warning to Humanity," statement and press release issued 18 November 1992 by the Union of Concerned Scientists.
14. United Nations Department of Information, "Population Pressures: A Complex Equation," *Earth Summit in Focus*, no. 6 (Febraury 1992).
15. Paul Kennedy, *Preparing for the Twenty-First Century* (New York: Random House, 1993), p. 33.
16. Nafis Sadik, Address to the Second Preparatory Committee, United Nations Conference on Environment and Development, 4 April 1991, Geneva.
17. Tim Wirth, "Nature, the Next Superpower," *Trilogy* 2(4) (November/December 1991): 20.

18. *Our Common Future: Report of the World Commission on Environment and Development* (New York: Oxford University Press, 1987).

2. The End of Innocence (pp. 12–28)

1. Larousse, *World Mythology* (Secaucus, N.J.: Chartwell Books, 1965), p. 68.
2. David Lowenthal, "Awareness of Human Impacts: Changing Attitudes and Emphases, in *The Earth as Transformed by Human Action,* ed. B. L. Turner II et al. (Cambridge: Cambridge University Press and Clark University, 1990), p. 121.
3. Clarence J. Glacken, *Traces on the Rhodian Shore* (Berkeley and Los Angeles: University of California Press, 1967), p. 707.
4. Ibid., p. 5.
5. Gen. 1:28.
6. Rev. 7:3.
7. David Hume, *An Inquiry concerning the Principles of Morals,* quoted in *A Dictionary of Environmental Quotations,* ed. Barbara K. Rodes and Rice Odell (New York: Simon & Schuster, 1992), p. 196.
8. Jeremy Rifkin, *Biosphere Politics* (New York: Crown, 1991), p. 12.
9. Ibid., p. 122.
10. Jean Jacques Rousseau, *On the Inequality among Mankind,* vol. 24 of the Harvard Classics (New York: P. F. Collier & Son, 1909), p. 172.
11. Jean Jacques Rousseau, *The Fall from Nature,* quoted in *A Documentary History of Conservation in America,* ed. Robert McHenry with Charles Van Doren (New York: Praeger, 1972), pp. 21–22.
12. William Wordsworth, *The Tables Turned,* 1798.
13. Clive Ponting, *A Green History of the World: The Environment and the Collapse of Great Civilizations* (New York: Penguin Books, 1991), p. 87.
14. John Young, *Sustaining the Earth* (Cambridge, Mass.: Harvard University Press, 1990), p. 58.
15. Norman Myers, *Ultimate Security* (New York: W. W. Norton, 1993), p. 38.
16. Ponting, *A Green History,* p. 176.
17. Quoted in Rodes and Odell, *Dictionary of Environmental Quotations,* p. 144.
18. Fernand Braudel, *The Structures of Everyday Life,* vol. 1 of *Civilization and Capitalism* (New York: Harper & Row, 1981), p. 27.
19. Hayward R. Alker Jr. and Peter M. Haas, "The Rise of Global Ecopolitics," in *Global Accord,* ed. Nazli Choucri (Cambridge, Mass.: M.I.T. Press, 1993), p. 137.
20. Ponting, *A Green History,* p. 116.
21. Robert W. Kates, B. L. Turner II, and William C. Clark, "The Great Transformation," in B. L. Turner II et al., eds, *The Earth as Transformed by Human Action* (Cambridge: Cambridge University Press and Clark University, 1990), p. 1.
22. Ibid., p. 11.
23. Alker and Haas, "Rise of Global Ecopolitics," p. 163.
24. Roderick Nash, *The Rights of Nature* (Madison: University of Wisconsin Press, 1989), p. 55.

25. Donald Worster, *Nature's Economy* (Cambridge: Cambridge University Press, 1977), p. 127.
26. Charles Darwin, *On the Origin of Species* (New York: Heritage Press, 1963), p. 444.
27. Friedrich Engels, *Outlines of a Critique of Political Economy,* quoted in Rodes and Odell, *A Dictionary of Environmental Quotations,* p. 166.
28. Richard J. Barnet, *The Lean Years* (New York: Simon and Schuster, 1980).
29. Henry David Thoreau, *Walden* (London, J. M. Dent & Sons, 1930), p. 78.
30. George Perkins Marsh, *Man and Nature,* ed. David Lowenthal (Cambridge, Mass.: Harvard University Press, 1965), p. 36.
31. Ibid., p. 34.
32. Ibid., p. xxvii.
33. Ibid., p. 45.
34. Theodore Roszak, *The Voice of the Earth* (New York: Simon & Schuster, 1992), p. 186.
35. John McCormick, *Reclaiming Paradise: The Global Environmental Movement* (Bloomington: Indiana University Press, 1991), p. 5.
36. Ibid., p. 15.
37. Ibid., pp. 8–9.
38. Philip Shabecoff, *A Fierce Green Fire* (New York: Hill and Wang, 1993), p. 68.
39. Lynton Keith Caldwell, *International Environmental Policy*, 2d ed. (Durham, N.C.: Duke University Press, 1990), p. 32.
40. Edith Weiss Brown, "International Environmental Law: Contemporary Issues and the Emergence of a New World Order," *Georgetown Law Journal* 81 (March 1993): 675.
41. McCormick, *Reclaiming Paradise,* pp. 16–17.
42. Hans Huth, *Nature and the American* (Lincoln: University of Nebraska Press, 1990), pp. 209–10.
43. Caldwell, *International Environmental Policy,* p. 42.
44. Robert Boardman, *International Organization and the Conservation of Nature* (Bloomington: Indiana University Press, 1981), p. 27.
45. Ibid.
46. Caldwell, *International Environmental Policy,* p. 34.
47. Brown, "International Environmental Law," p. 676.
48. Andy Crump, *Dictionary of Environment and Development* (Cambridge, Mass.: MIT Press, 1993), p. 140
49. Boardman, *International Organization,* p. 31.
50. Alker and Haas, "Rise of Global Ecopolitics," pp. 140–41.
51. Kenneth N. Townsend, "Steady State Economies and the Command Economy," in *Valuing the Earth,* ed. Herman E. Daly and Kenneth N. Townsend (Cambridge, Mass.: M.I.T. Press, 1993), p. 277.
52. Quoted in Caldwell, *International Environmental Policy,* p. 25.
53. John Passmore, *Man's Responsibility for Nature* (New York: Charles Scribner's Sons, 1974), p. 34.

54. Quoted in Roderick Frazier Nash, *The Rights of Nature* (Madison: University of Wisconsin Press, 1989), pp. 60–61.
55. Lewis Mumford, *Techniques and Civilization* (New York: Harcourt, Brace & World, 1962).
56. Aldo Leopold, *A Sand County Almanac* (New York: Oxford University Press, 1948), p. 215.
57. Ibid.
58. Worster, *Nature's Economy*, p. 339.
59. B. L. Turner et al., *The Earth as Transformed by Human Action* (Cambridge: Cambridge University Press and Clark University, 1990) p. 12.
60. Milton Russell, "Environmental Protection for the 1990s and Beyond," *Environment* 29 (7) (September 1987).
61. Barry Commoner, *The Closing Circle* (New York: Alfred A. Knopf, 1971), p. 299.
62. Donella H. Meadows, Dennis L. Meadows, Jørgen Randers, and William W. Behrens III, *The Limits to Growth* (New York: Universe Books, 1972), pp. 23–24.
63. Ibid., p. 194.
64. Lowenthal, "Awareness of Human Impacts," in Turner et al., *The Earth as Transformed*, p. 129.
65. Rachel Carson, *Silent Spring*, 25th anniversary ed. (Boston: Houghton Mifflin, 1987).
66. Ibid., p. 6.
67. Andrew Jamison, Ron Eyerman, and Jacqueline Cramer, *The Making of the New Environmental Consciousness* (Edinburgh: Edinburgh University Press, 1990), p. 19.
68. U Thant, speech delivered at the University of Texas, Austin, Tex., 14 May 1970, quoted in Rodes and Odell, *A Dictionary of Environmental Quotations*, p. 163.

3. Springtime in Stockholm (pp. 29–43)

1. Quoted in Brian Urquhart, *A Life in Peace and War* (New York: Harper & Row, 1987), p. 99.
2. Ibid., p. 103.
3. Caroline Thomas, *The Environment in International Relations* (London: Royal Institute of International Affairs, 1992), pp. 4–5.
4. Ibid., p. 7.
5. Lynton Keith Caldwell, *International Environmental Policy*, ed ed. (Durham, N.C.: Duke University Press, 1990), p. 42.
6. John McCormick, *Reclaiming Paradise: The Global Environmental Movement* (Bloomington: Indiana University Press, 1991), p. 60.
7. Ibid., pp. 88–90.
8. Caldwell, *International Environmental Policy,* p. 46.
9. McCormick, *Reclaiming Paradise,* p. 182.
10. Peter Stone, *Did We Save the Earth at Stockholm?* (London: Earth Island, 1973), p. 18.

11. Ibid.
12. Except as otherwise noted, biographical information about Maurice Strong is from an interview with Mr. Strong tape-recorded by the author in Toronto over a two-day period in October 1992.
13. Daniel Wood, "The Wizard of Baca Grande," *West,* May 1990, p. 38.
14. Alison Carper, "Strong Tactics," *Los Angeles Times Magazine,* 17 May 1992.
15. Gareth Porter and Janet Welsh Brown, *Global Environmental Politics* (Boulder, Colo.: Westview Press, 1991), p. 45.
16. McCormick, *Reclaiming Paradise,* p. 91.
17. Stone, *Did We Save the Earth,* p. 28.
18. "Report Submitted by a Panel of Experts Convened by the Secretary General of the United Nations Conference on the Human Environment, 4–12 June 1971."
19. Strong, interview.
20. Maurice F. Strong, *From Stockholm to Rio: A Journey Down a Generation,* Earth Summit Publication no. 1 (New York: United Nations Conference on Environment and Development, 1991), p. 1.
21. Some accounts put the number of nations sending delegates to Stockholm at 114, but Strong, in subsequent writings and in his interview with the author, put the number at 113.
22. "Brazil, U.K., et al.: Only 113 Earths," *Stockholm Conference Eco,* 9 June 1972, p. 1.
23. Walter Sullivan, "Struggling against the Doomsday Timetable," *New York Times,* 12 June 1972, "News of the Week in Review."
24. Claire Sterling, "Pollution Meeting Hails Gandhi Talk," *Washington Post,* 15 June 1972.
25. Anthony Lewis, "One Confused Earth," *The New York Times,* 17 June 1972, "News of the Week in Review."
26. Ibid.
27. Hans Landsberg, "Reflections on the Stockholm Conference," (unpublished report to Resources for the Future), 1972, p. 3.
28. The Chinese delegation would not vote in favor of the declaration because of the nuclear weapons provision, but Strong persuaded the Chinese to agree to be enrolled as present but not voting in order to have no objections to the statement.
29. Louis B. Sohn, "The Stockholm Declaration on the Human Environment," *Harvard International Law Journal 14* (Summer 1973): 513–15.
30. Peter M. Haas, Robert O. Keohane, and Marc A. Levy, eds., *Institutions for the Earth* (Cambridge, Mass.: M.I.T. press, 1993), p. 6.
31. Landsberg, "Reflections," p. 19.
32. "United Nations Conference on the Human Environment, Resolution on Institutional and Financial Arrangements," in Stone, *Did We Save the Earth*, pp. 192–94.
33. Nazli Choucri, ed., *Global Accord* (Cambridge, Mass.: M.I.T. Press, 1993), p. 151.
34. Maurice Strong, "One Year after Stockholm: An Ecological Approach to Management," *Foreign Affairs* (July 1973): 692.
35. "The Need to Work in Concert Was Clear," *New York Times* 18 June 1972, "News of the Week in Review."

36. Thomas, *Environment in International Relations*, p. 26.
37. Caldwell, *International Environmental Policy,* p. 83.

4. The World As It Is (pp. 44–58)

1. Maurice F. Strong, *From Stockholm to Rio: A Journey Down a Generation*, Earth Summit Publication (New York: United Nations Conference on Economic Development, 1991), p. 2.
2. Quoted in "State of the Environment: The Bad News, The Good News," UNEP news release, 11 December 1992.
3. Donella H. Meadows, Dennis L. Meadows, and Jorgen Randers, *Beyond the Limits* (Post Mills, Vt.: Chelsea Green, 1992), p. 2.
4. Philip Shabecoff, "U.N. Parley Will Examine How Environment Has Done in the Last 10 Years," *New York Times,* 3 May 1982.
5. Ibid.
6. Quoted by William K. Reilly in "Risky Business: Life, Death Pollution and the Global Environment" (paper presented to the Institute for International Studies.)
7. John McCormick, *Reclaiming Paradise: The Global Environmental Movement* (Bloomington: Indiana University Press, 1991), pp. 154–57.
8. Lynton Keith Caldwell, Appendix C, "Selected Treaties of Environmental Significance," in *International Environmental Policy*, 2nd ed. (Durham, N.C.: Duke University Press, 1991), pp. 349–51.
9. Peter S. Thacher, "Background to Institutional Options for Management of the Global Environment and Commons," preliminary paper for the World Federation of United Nations Associations Project, "Global Security and Risk Management," p. 2.
10. *UNEP Profile* (Nairobi, Kenya: United Nations Environment Programme, 1990).
11. Quoted in Erik P. Eckholm, *Down to Earth* (New York: W. W. Norton, 1982), p. 89.
12. See, for example, McCormick, *Reclaiming Paradise,* p. 51.]
13. Gareth Porter and Janet Welsh Brown, *Global Environmental Politics* (Boulder, Colo.: Westview Press, 1991), p. 49.
14. Mostafa K. Tolba, interview by author, September 1992.
15. Brian Urquhart, interview by author, March 1992.
16. Russell Train, interview by author, 1982.
17. Lew Rockwell, "An Anti-Environmentalist Manifesto," *Patrick J. Buchanan From the Right* 1(6) (1990): 1–2.
18. Sir Crispin Tickell, "The Inevitability of Environmental Security," in *Threats without Enemies,* ed. Gwyn Prins (London: Earthscan Publications, 1993), p. 17.
19. Boutros Boutros-Ghali, "Somalia's Cry" (speech presented at photographic exhibition, 16 December 1992), quoted in *Go-Between* (United Nations Liaison Service), December 1992.
20. Lester R. Brown, Christopher Flavin, and Hal Kane, *Vital Signs 1992* (W. W. Norton, 1992), p. 85.
21. Norman Myers, *Ultimate Security* (New York: W. W. Norton, 1993), pp. 219–20.

Myers cites R. L. Sivard, *World Military and Social Expenditures 1991* (Washington, D.C.: World Priorities, 1991).

22. Porter and Brown, *Global Environmental Politics*, p. 115.
23. Jeremy Rifkin, *Biosphere Politics* (New York: Crown, 1991) p. 143.
24. Adam Swift, *Global Political Ecology* (London: Pluto Press, 1993), p. 55.
25. *Challenge to the South: An Overview and Summary of the South Commission Report* (Geneva: South Commission, 1990), p. 60.
26. *World Resources 1992–93* (New York: Oxford University Press, 1992), p. 76.
27. Ibid., p. 246.
28. *Challenge to the South,* p. 61.
29. *The WIT Report* (World Information Transfer) 4(2) (March/April 1992): 12.
30. *World Resources 1992–93,* p. 242.
31. Shridath Ramphal, *Our Country the Planet* (Washington, D.C.: Island Press, 1992), p. 188.
32. Ibid., p. 184.
33. Quoted by Peter Raven in "Carrying Capacity of the Globe: What Kind of World Do We Want?" (paper presented to Sigma Xi conference on Global Change and the Human Prospect, November 17, 1991).
34. Sandra Postel, "Carrying Capacity: Earth's Bottom Line," in *State of the World 1994,* ed. Lester Brown (New York: W. W. Norton, 1994), p. 5.
35. *Consumption Patterns: The Driving Force of Environmental Stress* (Bombay: Indira Ghandi Institute of Development Research, 1991), pp. 1–4.
36. Adam Swift, *Global Political Ecology*, p. 171.
37. Nazli Choucri, "Multinational Corporations and the Global Environment," in *Global Accord* (Cambridge, Mass.: M.I.T. Press, 1993), p. 247.
38. Alan Durning, *How Much Is Enough?* (New York: W. W. Norton, 1992), p. 23.
39. Eckholm, *Down to Earth*, p. xii.
40. Louise Tappa, "The Theology of Environment and Development," in *SONED on UNCED, A Southern Perspective on the Environment and Development Crisis* (Geneva: Southern Networks for Development, African Region, 1991), p. 64.
41. *Challenge to the South,* p. 5.
42. Adam Swift, *Global Political Economy,* p. 31.
43. Barry Commoner, *Making Peace with the Planet* (New York: Pantheon, 1990), p. 52.
44. Global Tomorrow Coalition, *The Global Ecology Handbook* (Boston: Beacon Press, 1990), p. 253.
45. Helen Caldicott, M.D., *If You Love This Planet* (New York: W. W. Norton, 1992), p. 92.
46. Paul R. Ehrlich and Anne H. Ehrlich, *Healing the Planet* (Reading, Mass.: Addison-Wesley, 1991), p. 188.
47. Mikhail Kokeev, interview by author, 1991.
48. For an excellent firsthand account of the environmental collapse of eastern Europe, see Larry Tye, *Boston Globe*, 17–21 December 1989.
49. Robert S. McNamara, "Africa's Development Crisis: Agricultural Stagnation,

Population Explosion and Environmental Degradation" (paper presented to the African Leadership Forum, Nigeria, 21 June 1990).

50. Lester R. Brown, Hal Kane, and Ed Ayres, *Vital Signs 1993* (New York: W. W. Norton, 1993), p. 41.
51. Ibid., p. 52.
52. Thacher, "Background to Institutional Options," p. 1.
53. Eric Chivian, Michael McCally, Howard Hu, and Andrew Haines, *Critical Condition: Human Health and the Environment* (Cambridge, Mass.: M.I.T. Press, 1993), pp. ix, x.
54. Porter and Brown, *Global Environmental Politics,* p. 28.
55. Caldwell, *International Environmetal Policy*, p. 210.
56. Jessica Tuchman Mathews, "Nations and Nature: A New View of Security," in *Threats without Enemies*, ed. Gwyn Prins (London: Earthscan Publications, 1993), p. 25.
57. Mark Sagoff, "What Is Environmentalism?" (paper prepared for the National Commission on the Environment, 1991), p. 32.

5. Fire from Below (pp. 62–76)

1. Philip Shabecoff, *A Fierce Green Fire* (New York: Hill and Wang, 1993), p. 69.
2. David Vogel, "Environmental Policy in Europe and Japan," in *Environmental Policy in the 1990s*, ed. Norman J. Vig and Michael E. Kraft (Washington, D.C.: CQ Press, 1990), p. 258.
3. Enrique Iglesias, speech to UN Conference on Local Government, New York, December 1989.
4. Riley E. Dunlap, George H. Gallup Jr., and Alec M. Gallup, "Of Global Concern: Results of the Health of the Planet Survey, *Environment* 35(9) (November 1993): 7.
5. *Defending the Earth: Abuses of Human Rights and the Environment* report of the Human Rights Watch of the Natural Resources Defense Council, June 1992 (New York: NRDC Publishers, 1992), p. v.
6. Ibid., p. vi.
7. Alan B. Durning, *Action at the Grassroots: Fighting Poverty and Environmental Decline,* Worldwatch Paper 88 (Washington, D.C.: Worldwatch Institute, 1989), p. 13.
8. *Defending the Earth,* p. 42.
9. Quoted by Michael J. Kane in *International Protection of Human Rights and the Environment* (Washington, D.C.: U.S. Environmental Protection Agency, 1991).
10. Michael McCloskey, "The Emerging Worldwide Environmental Movement" (paper presented to the Western Public Interest Law Conference, 5 March 1988).
11. Andy Crump, *Dictionary of Environment and Development* (Cambridge, Mass.: MIT Press. 1993), p. 267.
12. *Caring for the Earth: A Strategy for Sustainable Living* (Gland, Switzerland: IUCN, UNEP, WWF, 1991), pp. 1–4.
13. Durning, *Action at the Grassroots,* p. 7.

14. Gareth Porter and Janet Welsh Brown, *Global Environmental Politics,* (Boulder, Colo.: Westview Press, 1991), p. 57.
15. See, e.g., Susan Griffin, *Woman and Nature* (New York: Harper Colophon, 1978).
16. Charlene Spretnak and Fritjof Capra, *Green Politics* (Santa Fe, N.M.: Bear & Co., 1986), p. 53.
17. Jodi L. Jacobson, *Gender Bias: Roadblock to Sustainable Development,* Worldwatch Paper 110 (Washington, D.C.: Worldwatch Institute, 1992).
18. Perdita Huston, "Why All the Fuss about Women?" *Crosscurrents,* no. 3 (9–11 March 1992): 13.
19. Lynn White Jr., "The Historical Roots of Our Ecological Crisis," *Science* 155 (10 March 1967): 1203–7.
20. Jan-Olaf Willums and Ulrich Goluke, *From Ideas to Action: Business and Sustainable Development* (Oslo: International Environment Bureau of the ICC, 1992), p. 83.
21. Ibid., p. 13.
22. *Independent Sectors Network '92* no. 5 (March 1991): 7.
23. Peter H. Sand, "International Cooperation: The Environmental Experience," in *Preserving the Global Environment* ed. Jessica Tuchman Mathews (New York: W. W. Norton, 1991), p. 267.
24. Horst Mewes, "The Green Party Comes of Age," *Environment* 27(5) (June 1985): 14.
25. Spretnak and Capra, *Green Politics,* pp. 165–71.
26. Wolfgang Rudig, "Green Party Politics around the World," *Environment* (October 1991): 7.
27. John McCormick, *Reclaiming Paradise: The Global Environmental Movement* (Bloomington: Indiana University Press, 1991), p. 140
28. Spretnak and Capra, *Green Politics,* p. 50.
29. For a discussion of the role of ecology in Nazi Germany, see Anna Bramwell, *Ecology in the 20th Century* (New Haven, Conn.: Yale University Press, 1989.)
30. Jessica Tuchman Mathews, "Nations and Nature: A New View of Security," in *Threats Without Enemies,* ed. Gwyn Prins (London: Earthscan Publications, 1993), p. 19.
31. William D. Montalbano, "Green Wave Surging over West Europe," *Los Angeles Times,* 11 May 1989.
32. Lester Brown, *State of the World 1991* (New York: W. W. Norton, 1991).
33. Montalbano, *Green Wave.*
34. These election results were obtained from an interview with Eric Odel, a staff member of the National Clearing House of Greens/Green Party USA.
35. Merry Ann Moore, "Ripe for a Change," *E* magazine September 1992.
36. Jonathan Porritt, *Seeing Green* (Oxford: Basil Blackford, 1984), p. 224.
37. Thomas, *Environment in International Relations,* p. 18.
38. Porter and Brown, *Global Environmental Politics,* p. 61.
39. UN Development Programme, "NGO Perspectives on Poverty, Environment and Development," New York, November 1991, pp. 19–20.
40. Barbara Jancar, "Chaos as an Explanation of the Role of Environmental Groups in

East European Politics," in *Green Politics Two,* ed. Wolfgang Rudig (Edinburgh: Edinburgh University Press, 1991), p. 182.

41. Hira Jhamtani, "The Imperialism of Northern NGOs," *Earth Island Journal,* Summer 1992, p. 5.
42. Richard J. Barnet and John Cavanagh, *Global Dreams* (New York: Simon & Schuster, 1994), p. 35.
43. ELCI presentation to the UN Conference on Environment and Development, Rio de Janeiro, June 1992, p. 1.
44. "Alliance Campaign News," *Northern Alliance,* December 1991, p. 3.
45. S. M. Mohamed Idris, "What Green Really Means for the Third World," *Third World Resurgence* no. 1 (spring 1990): 2.
46. *Environment Liaison Centre International,* "Justice Between Peoples—Justice Between Generations," prepared for Roots of the Future Conference, Paris, 17–29 December 1991.
47. Lester W. Milbrath, *Environmentalists: Vanguard for a New Society* (Albany: State University of New York Press, 1984).
48. Lynton Keith Caldwell, *International Economic Policy,* 2d ed. (Durham, N.C.: Duke University Press, 1990), p. 313.

6. (Political) Scientists (pp. 78–91)

1. Philip Shabecoff, "The Heat Is On: Calculating the Consequences of a Warmer Planet Earth," *New York Times,* 26 June 1988, "News of the Week in Review."
2. Michael Oppenheimer and Robert H. Boyle, *Dead Heat* (New York: Basic Books, 1990), p. 34.
3. Ibid., p. 40.
4. Stephen H. Schneider, *Global Warming* (San Francisco: Sierra Club Books, 1989), p. 211.
5. James H. Hansen, interview by author, June 1994.
6. Patrick J. Michaels, *Sound and Fury* (Washington, D.C.: Cato Institute, 1992), pp. 16–19. (The Cato Institute is sometimes referred to as a "right-wing think tank," but engaging in research designed solely to support the *a priori* conclusion that government activity is sinister and must be limited can hardly be called thinking.)
7. Schneider, *Global Warming,* p. 196.
8. Hansen interview.
9. Richard A. Kerr, "Hansen vs. the World on the Greenhouse Threat," *Research News,* 3 June 1989, p. 1041.
10. Gerald Piel and Osborn Sergerberg Jr., eds., *The World of René Dubos* (New York: Henry Holt, 1990), p. 219.
11. Ibid., p. 237.
12. Thomas Love Peacock, *Gryll Grange,* 1864, quoted in Rodes and Odell, *Dictionary of Environmental Quotations,* p. 82.
13. Stephen F. Mason, *A History of the Sciences* (New York: Collier Books, 1962), p. 601.

14. Oppenheimer and Boyle, *Dead Heat,* p. 43.
15. Eugene B. Skolnikoff, "Science and Technology: The Sources of Change," in *Global Accord,* ed. Nazli Choucri (Cambridge, Mass.: M.I.T. Press, 1993), p. 255.
16. Piel and Segerberg, *World of René Dubos,* p. 227.
17. Chellis Glendenning, *When Technology Wounds* (New York: William Morris, 1990), p. 20.
18. Silvio O. Funtowicz and Jerome R. Ravetz, "A New Scientific Methodology for Global Environmental Issues," in *Ecological Economics: The Science and Management of Sustainability,* ed. Robert Constanza (New York: Columbia University Press, 1991), pp. 137–38.
19. Anna Bramwell, *Ecology in the 20th Century: A History* (New Haven, Conn.: Yale University Press, 1989), p. 248.
20. See, for example, Ben Bolchman and Harold Lyons, *Apocalypse Not* (Washington, D.C.: Cato Institute, 1993).
21. Philip Shabecoff, *A Fierce Green Fire* (New York: Hill and Wang, 1993), p. 135.
22. Richard Elliott Benedick, *Ozone Diplomacy* (Cambridge, Mass.: Harvard University Press, 1991), p. 9.
23. Robert C. Paehlke, *Environmentalism and the Future of Progressive Politics* (New Haven, Conn.: Yale University Press, 1989), pp. 117–18.
24. Gregg Easterbrook, "Science and the Environment," in *Gale Environmental Almanac*, ed. Russ Hoyle (Detroit: Gale Research, 1993), p. 309.
25. Cheryl Simon Silver, with Ruth S. DeFries, *One Earth One Future* (Washington, D.C.: National Academy Press, 1990), pp. 15–16.
26. Ibid., p. 16.
27. Lynton Keith Caldwell, *International Economic Policy,* 2d ed. (Durham, N.C.: Duke University Press, 1990), p. 6.
28. Gary D. Brewer, "Environmental Challenges and Managerial Responses: A Decision Perspective," in *Global Accord*, ed. Nazli Choucri (Cambridge, Mass.: M.I.T. Press, 1993), p. 288.
29. John Flynn, "The Subversive Scientists," *Amicus Journal,* Fall 1990, pp. 4–9.
30. Ibid., p. 9.
31. George Perkins Marsh, *Man and Nature,* ed. David Lowenthal (Cambridge, Mass.: Belknap Press of Harvard University Press, 1965), p. 465.
32. Silver and DeFries, *One Earth One Future,* p. 2.
33. Oppenheimer and Boyle, *Dead Heat,* pp. 60–61.
34. Quoted in Oppenheimer and Boyle, *Dead Heat,* p. 44.
35. Tyler Prize brochure, 1983.
36. *Who's Who in Science in Europe,* 7th ed. (Essex, U.K.: Longman Group).
37. Robert Pool, "Struggling to Do Science for Society," *Science* 248 (11 May 1990): 672.
38. Easterbrook, "Science and the Environment," p. 327.
39. Funtowicz and Ravetz, "A New Scientific Methodology," p. 151.
40. Ibid.
41. Kai N. Lee, *Compass and Gyroscope* (Washington, D.C.: Island Press, 1993).
42. Milton Russell, "Environmental Protection: Laying the Foundation for the Year

2000," (paper presented to the Industrial Research Institute, Washington, D.C., 22 October 1985).

7. Getting and Spending (pp. 92–111)

1. Hazel Henderson, *The Politics of the Solar Age: Alternatives to Economics* (Indianapolis, Ind.: Knowledge Systems, 1988), p. 189.
2. E. F. Schumacher, *Small Is Beautiful* (New York: Harper & Row, 1973), p. 33.
3. Lester R. Brown, *Building a Sustainable Society* (New York: W. W. Norton, 1981).
4. World Commission on Environment and Development, *Our Common Future* (New York: Oxford University Press, 1987).
5. Ibid., p. ix.
6. Ibid., p. 4.
7. Ibid., p. 43.
8. Ibid.
9. Edith Brown Weiss, "Intergenerational Equity: Toward an International Legal Framework," in *Global Accord,* ed. Nazli Choucri (Cambridge, Mass.: M.I.T. Press, 1993), p. 334.
10. David Pearce, Anil Markandya, and Edward B. Barbier, *Blueprint for a Green Economy* (London: Earthscan International, 1989), p. xv.
11. World Commission on Environment and Development, *Our Common Future,* p. 52.
12. Ibid., p. 3.
13. See, for example, Adam Swift, *Global Political Ecology* (London: Pluto Press, 1993), p. 46.
14. Ibid., pp. 2–3. "Adam Swift" is identified on the book jacket as "the pseudonym of a former civil servant and senior official. He now represents an international non-governmental organization at the United Nations."
15. Herman E. Daly and Kenneth N. Townsend, introduction to *Valuing the Earth* (Cambridge, Mass.: M.I.T. Press, 1993), pp. 1–15.
16. Talbot Page, "Sustainability and the Problem of Valuation," in *Ecological Economics* ed. Robert Costanza, (New York: Columbia University Press, 1991), p. 64.
17. Kenneth Boulding, "The Economics of the Coming Spaceship Earth," in Daly and Townsend, *Valuing the Earth,* p. 303.
18. Ibid., p. 304.
19. Herman E. Daly, "Boundless Bull," *Gannett Center Journal,* summer 1990, p. 113.
20. Ibid., pp. 114–15.
21. Robert Costanza, Herman E. Daly, and Joy A. Bartholomew, "Goals, Agenda, and Policy Recommendations for Ecological Economics," in Costanza, *Ecological Economics,* pp. 3–7.
22. Ibid.
23. Jim MacNeill, *The Brundtland Commission: A Personal Reflection,* p. 5.
24. Ibid., p. 6.
25. Herman E. Daly, "Sustainable Growth: An Impossibility Theorem," in Daly and Townsend, *Valuing the Earth,* p. 267.

26. P. M. Vitousek, P. R. Ehrlich, A. H. Ehrlich, and P. M. Matson, "Human Appropriation of the Products of Photosynthesis" *BioScience,* 36(6) (1986): 368–73.
27. Daly, "Sustainable Growth," p. 269.
28. Daily and Townsend, *Valuing the Earth,* pp. 29–40.
29. Herman Daly, interview by author, 1991.
30. Nicholas Georgescu-Roegen, "Selections from 'Energy and Economic Myths,'" in Daly and Townsend, *Valuing the Earth,* p. 95.
31. Ibid., p. 81.
32. Juan Martinez-Alier, *Ecological Economics* (Oxford: Basil Blackwell, 1987), p. 132.
33. Gerald Alonzo Smith, "The Purpose of Wealth: A Historical Perspective," in Daly and Townsend, *Valuing the Earth,* p. 185.
34. J. A. Lutzenberger, "Keynote Address to International Meeting of Parliamentarians," Washington, D.C., 30 April 1990.
35. Peter A. Victor, Edward Hanna, and Atif Kubursi, "How Strong Is Weak Sustainability?" (paper presented at an International Symposium on Models of Sustainable Development sponsored by the Université Pantheon-Sorbonne and the Association francaise des science et technologies de l'information et des systemes, Paris, 16–18 March 1994, Vol. I., p. 99.
36. Ibid., p. 96.
37. Pearce et al., *Blueprint,* p. 22.
38. Costanza et al., "Goals," p. 11.
39. Salah El Serafy, remarks at seminar, Models of Sustainable Development, Paris, 16 March 1994.
40. A. C. Pigou, *The Economics of Welfare* (1920; reprint, London: Macmillan, 1952).
41. David Pearce, *Economic Values and the Natural World* (Cambridge, Mass.: M.I.T. Press, 1993), pp. 93–94.
42. National Commission on the Environment, *Choosing a Sustainable Future* (Washington, D.C.: Island Press, 1993), p. 26.
43. Ibid.
44. Ernst U. von Weizacker and Jochen Jesinghaus, *Ecological Tax Reform* (London: Zed Books, 1993).
45. See, for example, Philip Shabecoff, "World Lenders Facing Pressure from Ecologists," *New York Times,* 30 October 1986.
46. Roger D. Stone and Eve Hamilton, *Global Economics and the Environment* (New York: Council on Foreign Relations Press, 1991), p. 38.
47. Shabecoff, "World Lenders."
48. Ibid.
49. Mohamed El-Ashry, interview by author, 1991.
50. *The World Bank and the Environment: A Progress Report* (Washington, D.C.: World Bank, 1992), pp. 1–2.
51. Stone and Hamilton, *Global Economics*, p. 36.
52. Hillary French, *Costly Tradeoffs: Reconciling Trade and the Environment* (Washington, D.C.: Worldwatch Institute, 1993), p. 6.

53. *The GATT Report on Trade and Environment: A Critique by the World Wide Fund for Nature* (Gland, Switzerland: WWFN, 1992
54. French, *Costly Tradeoffs,* p. 5.
55. Pete Nelson, "GATT: Enviros May Switch from Pro-NAFTA to Anti-GATT," *Greenwire,* 16 November 1993, p. 3.
56. Edith Brown Weiss, "Environment and Trade as Partners in Sustainable Development: A Commentary," *American Journal of International Law* 86 (October 1992): 728.
57. Ibid., p. 735.
58. Swift, *Global Politics Ecology*, pp. 172–73.
59. Henderson, *Political of the Solar Age,* p. 103.
60. Nicholas L. Reding, "The Structure of Environmental Revolution: The Industrial Paradigm Is Shifting," Fontbonne College Lecture, St. Louis, 13 November 1990.
61. Linda Starke, *Signs of Hope* (New York: Oxford University Press, 1990), p. 103.
62. Stephen Schmidheiny, with the Business Council on Sustainable Development, *Changing Course* (Cambridge, Mass.: M.I.T. Press, 1992).
63. Stephen Schmidheiny, "Sustainable Development: A Global Challenge for Industry" (paper presented to the Environment and Development Seminar, Rio de Janeiro, 7–8 March 1991.
64. Philip Shabecoff, *A Fierce Green Fire* (New York: Hill and Wang, 1993), p. 145.
65. Martinez-Alier, *Ecological Economics*, p. xxv.
66. Robert W. Fri, "Expanding Economics," *Economy and Environment* 3(1) (spring 1994): 2.
67. Mark Sagoff, "On the Expansion of Economic Theory: A Rejoinder," *Economy and Environment* 3(2) (summer 1994): 7.

8. The Greening of Geopolitics (pp. 112–137)

1. Philip Shabecoff, "Dozens of Nations Approve Accord to Protect Ozone, *New York Times*, 17 September 1987.
2. Quoted in Richard Elliot Benedick, *Ozone Diplomacy* (Cambridge, Mass.: Harvard University Press, 1991), p. 210.
3. Ibid.
4. Nazli Choucri, *Global Accord* (Cambridge, Mass.: M.I.T. Press, 1993), p. 405.
5. William Wertenbaker, "A Reporter at Large: The Law of the Sea–1," *New Yorker,* 1 August 1983.
6. Ibid.
7. Lynton Keith Caldwell, *International Environmental Policy,* 2d ed. (Durham, N.C.: Duke University Press, 1991), p. 329.
8. Mikhail S. Gorbachev, speech to the UN General Assembly, New York, 7 December 1988.
9. Philip Shabecoff, "Suddenly, the World Itself Is a World Issue," *New York Times,* 25 December 1988, "News of the Week in Review."
10. Ibid.

11. Communique text in *New York Times,* 17 July 1989.
12. Nathan Gardels, "The Greening of International Affairs," interview with Gro Harlem Brundtland, *San Francisco Chronicle,* 12 April 1989.
13. Ibid.
14. Gwyn Prins, "Politics and the Environment," in *Threats without Enemies,* (London: Earthscan Publications, 1993), p. 184.
15. Henry Kissinger, *Diplomacy* (New York: Simon & Schuster, 1994), p. 24.
16. Ibid., pp. 805–34.
17. Quoted in Philip Shabecoff, "Traditional Definitions of National Security Are Shaken by Global Environmental Threats," *New York Times,* 29 May 1989.
18. Norman Myers, *Ultimate Security* (New York: W. W. Norton, 1993), p. 231.
19. Thomas F. Homer-Dixon, *Environmental Change and Violent Conflict,* Occasional Paper no. 4, International Security Studies Program (New York: American Academy of Arts and Sciences, 1990), p. 6.
20. Ibid.
21. Myers, *Ultimate Security,* p. 38.
22. Senator Sam Nunn, "Strategic Environmental Research Program" (speech, before the Senate, Washington, D.C., 28 June 1990).
23. Admiral Sir Julian Oswald, "Accepting the Challenge of Environmental Security," in *Threats without Enemies,* ed. Gwyn Prins (London: Earthscan Publications, 1993), pp. 115–18.
24. Sir Crispin Tickell, "Grasping the Concept of Environmental Security," in Pryns, *Threats without Enemies,* p. 23.
25. Senator Al Gore, (paper presented to Forum on Global Change and Our Common Future, National Academy of Sciences, Washington, D.C., 1 May 1989).
26. *Environmental Security: A DOD Partnership for Peace,* Strategic Studies Institute Special Report, ed. Kent Hughes Butts, (Washington, D.C.: U.S. Army War College, 1994), p. 1–2.
27. See, *inter alia,* Gareth Porter and Janet Welsh Brown, *Global Environmental Politics* (Boulder, Colo.: Westview Press, 1991); Jessica Tuchman Mathews, "Nations and Nature: A New View of Security," in *Threats without Enemies;* ed. Gwyn Prins (London: Earthscan Publications, 1993), and Norman Myers, *Ultimate Security.*
28. "The CSE Statement on Global Environmental Democracy," Centre for Science and Environment (undated), p. 1.
29. Porter and Brown, *Global Environmental Politics,* p. 127.
30. Lawrence E. Susskind, *Environmental Diplomacy* (New York: Oxford University Press, 1994), pp. 18–19.
31. Emil Salim, "Parliamentarians with a Global Responsibility" (paper presented to the Interparliamentary Conference on the Global Environment, Washington, D.C., 1 May 1990).
32. Porter and Brown, *Global Environmental Politics,* p. 129.
33. Cheryl Simon Silver and Ruth S. DeFries, *One Earth One Future* (Washington, D.C.: National Academy Press, 1990), p. 53.
34. Gore, *Earth in the Balance*, 1992.
35. Susskind, *Environmental Diplomacy,* p. 21.

36. Editors of *Harvard Law Review, Trends in International Environmental Law* (Chicago: American Bar Assocation, Section of International Law and Practice, 1992), p. 10.
37. Nicholas A. Robinson, "Diplomacy's New Frontiers," in *Gale Environmental Almanac* (Detroit: Gale Research, 1993), pp. 418–19.
38. Margaret Brusasco-MacKenzie, interview by author, 1991.
39. Editors of *Harvard Law Review, Trends,* p. 81.
40. Peter H. Sand, *Lessons Learned in Global Governance* (Washington, D.C.: World Resources Institute, 1990), p. 26.
41. Nazli Choucri and Robert C. North, "Global Accord: Imperative for the Twenty-First Century," in Choucri, *Global Accord* (Cambridge, Mass.: M.I.T. Press, 1993), p. 492.
42. Porter and Brown, *Global Environmental Politics,* p. 134.
43. "A New Partnership for Development: The Cartagena Commitment," final declaration of the UN Conference on Trade and Development, sess. 8, 27 February 1992.
44. Ibid., p. 13.
45. Ibid., p. 14.
46. Caroline Thomas, *The Environment in International Affairs,* 1992, pp. 113–14.
47. "CSE Statement," pp. 7–8.
48. Steve Rayner, "A Cultural Perspective on the Structure and Implementation of Global Environmental Agreements," *Evaluation Review*, February 1991, p. 79.
49. Richard Elliot Benedick, *Ozone Diplomacy* (Cambridge, Mass.: Harvard University Press, 1991), p. 45.
50. Rayner, "Cultural Perspective," p. 80.
51. Bendick, *Ozone Diplomacy,* p. 5.
52. Fang Lizhi, *Bringing Down the Great Wall* (New York: Alfred A. Knopf, 1991), pp. xlii–xliii, quoted in Michael J. Kane, "Promoting Political Rights and Protecting the Environment," Washington D.C., April, 1992, unpublished.
53. Quoted in Caldwell, *International Environmental Policy,* p. 335.
54. Lynton K. Caldwell, "Globalizing Environmentalism: Threshold of a New Phase in International Relations," School of Public and Environmental Affairs, Bloomington: Indiana University, Aug. 10, 1990, p. 21.
55. Pryns, "Politics and the Environment," p. 183.
56. Hillary French, statement at news conference on international governance, New York, 13 March 1992.

9. Slouching toward Rio (pp. 128–141)

1. Information supplied by Mikhail Kokeev, director of environmental affairs in the Soviet foreign ministry, interview by author, August 1991.
2. Maurice Strong, interview by author, March 1991.
3. *Brundtland Bulletin* (Centre for Our Common Future), March 1990, p. 2.
4. Strong interview.

5. May Davison, special assistant to Maurice Strong, interview by author, March 1991.
6. Strong interview.
7. Ibid.
8. Ibid.
9. Background interview by author.
10. Lance Antrim, "Report on the Processes of Negotiation at the Preparatory Committee of the U.N. Conference on Environment and Development," working paper, Project on Multilateral Negotiation of the American Academy of Diplomacy and the Paul H. Nitze School of Advanced International Studies, Johns Hopkins University, Baltimore, p. 27.
11. Antrim, "Report," p. 22.
12. Barbara Bramble, director of international activities for the National Wildlife Federation, interview by author, September 1992.
13. Strong interview.
14. Transcript of statement by Dr. Koofi Awoonor, to the second session of the Preparatory Committee, Geneva, 3 April 1991, pp. 2–4.
15. Unless otherwise indicated, all descriptions of and quotations from the preparatory committee meetings are from the author's notes taken at those meetings.
16. Remarks at an Earth Summit media briefing, a project of Island Press/Center for Resource Economics, Washington, D.C., 10 January 1992.
17. *Guidelines for the U.N. Environmental Conference: A United Nations Assessment Project Study* (Washington, D.C.: Heritage Foundation, 1991), pp. 1–2.
18. Conversation with the author.
19. Quotes and descriptions of Strong's meetings in Tokyo are from the author's notes, taken at the meetings.
20. Conversation with the author, 1991.

10. To the Wire (pp. 144–158)

1. Quotes from and descriptions of the preparatory committee meeting are from the author's notes, except where otherwise noted.
2. *The Global Partnership for Environment and Development: A Guide to Agenda 21* (Geneva: UN Conference on Environment and Development, 1992).
3. Maurice Strong, interview by author, October 1992.
4. Paul Lewis, "Environment Aid for Poor Nations Agreed at the U.N.," *New York Times,* 5 April 1992.
5. Marker press conference, 16 March 1992.
6. Kokeev press conference, 2 March 1992.
7. Richard F. Shepard, "Adios Prepcom 4; May Your Aim Be Recycled," *Earth Summit Times,* 3 April 1992, p. 2.
8. Barbara Bramble, National Wildlife Federation, interview by author, March 1992.
9. Elizabeth Barratt-Brown, interview by author, October 1992.

10. Michael McCoy, "NGO Role Grows, But There's a Limit," *Earth Times,* 31 March 1992, p. 2.
11. Robert Ryan, interview by author, October 1992.
12. Rafe Pomerance, interview by author, October 1992.
13. "Earth Summit: Bush to Go; White House Still 'Recalcitrant'," *Greenwire,* 7 May 1982, p. 9.
14. "Biodiversity: Enviros Accuse U.S. of Weakening Treaty," *Greenwire,* 19 May 1992, p. 3.
15. Mostafa K. Tolba, interview by author, September 1992.
16. Paul Lewis, "U.S. under Fire in Talks at U.N. on Environment," *New York Times,* 24 March 1992.
17. "Koh to Plenary: Time Is Running Out," *Earth Summit Times,* 31 March 1992, p. 8.
18. Johannah Bernstein, Pamela Chasek, and Langston James Goree VI, "A Summary of the Proceedings of the Fourth Session of the UNCED Preparatory Committee," *Earth Summit Bulletin,* April 1992, p. 11.
19. "Proposal of the Chairman of the Preparatory Committee for UNCED on the Rio Declaration on Environment and Development," in *Global Partnership,* pp. 1–4.
20. Ibid.
21. Nitin Desai, interview by author.
22. Rose Gutfeld, "How Bush Achieved Global Warming Pact with Modest Goals," *Wall Street Journal,* 27 May 1992.
23. Michael Weisskopf, "U.S. View Prevails on Emissions," *Washington Post,* 9 May 1992.
24. Jim MacNeill, "The Unfinished Business of PrepCom 4," *Earth Summit Times,* 3 April 1992, p. 13.

11. At the Summit (pp. 160–177)

1. Peter M. Haas, Marc A. Levy, and Edward A. Parson, "Appraising the Earth Summit," *Environment,* October 1992, p. 7.
2. "Earth Summit: The Capital of the Planet Is Dismantled," *Terraviva,* 15 June 1992, p. 22.
3. Quotes from and descriptions of the conference are from the author's notes, unless otherwise noted.
4. Jim MacNeill, "Honesty, Courage Needed to Save the Summit," *Earth Summit Times,* 3 June 1992, p. 1.
5. David E. Pitt, "Lines Tighten for Summit," *Earth Summit Times,* 3 June 1992, p. 1.
6. "Biodiversity: U.S. Won't Sign On," *Greenwire,* 1 June 1992, p. 1.
7. "EC: Ripa di Meana Says He Won't Attend Summit," *Greenwire,* 28 May 1992, p. 3.
8. "US Singled Out as Eco Bad Guy," *Jornal Do Brasil,* 5 June 1992, p. 1.
9. David E. Pitt, "Europe Moves to Aid U.S. in Rio," *Earth Summit Times,* 6 June 1992, p. 1.

10. "Earth Summit: The United States and the New Isolationism," *Greenwire,* 8 June 1992, p. 3.
11. For an excellent account of the negotiations on financial resources at the summit, see "A Summary of the Proceedings of the United Nations Conference on Environment and Development, 3–14 June 1992," *Earth Summit Bulletin,* 16 June 1992, pp. 5–7.
12. Ibid., p. 9.
13. "Forests: Delegates Lean Towards Limited Pact," *Greenwire,* 9 June 1992, p. 3.
14. "Earth Summit: Forest Principles Being Diplomatically Clearcut," *Greenwire,* 12 June 1992, p. 3.
15. Ibid.
16. David E. Pitt and Noel L. Gerson, "The Summit's Late, Late, Late Show," *Earth Summit Times,* 11 June 1992, p. 1.
17. Text of Li Peng speech in *Terraviva,* 15 June 1992, p. 10.
18. Philip Shabecoff, "Real Rio," *Buzzworm,* September/October 1992, p. 40.
19. David E. Pitt, "Bush's Speech Leaves Summit Feeling Low," *Earth Summit Times,* 13 June 1993, p. 1.
20. Senator Al Gore, interview by author, June 1992.
21. *Wall Street Journal,* 15 June 1992.
22. Quoted in *Europe* magazine, June 1992.
23. Emil Salim, interview by author, June 1992.
24. Teodomiro Braga, " 'I'm Not President of the World' Says Bush," *Jornal do Brasil,* 14 June 1992, p. 5.
25. Maurice Strong, interview by author, October 1992.
26. Maurice Strong, conversation with author, June 1992.
27. "NGOs at UNCED and Its Parallel Events," *E & D File 1992,* (UN Non-Governmental Liaison Service) July 1992, p. 1.
28. Martin Khor, "Still Fighting for One World," *Crosscurrents* (an independent NGO newspaper for UNCED) June 15 1992.
29. All of Strong's comments are from interviews by the author, except where otherwise noted.
30. Jamsheed Marker, interview by author, November 1992.
31. Razali Ismail, interview by author, November 1992.
32. Gro Harlem Brundtland, interview by author.
33. Sir Crispin Tickell, "The Inevitability of Environmental Security," in *Threats Without Enemies,* ed. Gwyn Prins (London: Earthscan Publications, 1993), p. 23.
34. Lawrence E. Susskind, *Environmental Diplomacy* (New York: Oxford University Press, 1994), pp. 41–42.
35. Norman Myers, *Ultimate Security* (New York: W. W. Norton, 1993), p. 249.
36. Philip Shabecoff, "Assessing Rio: Where's the Churrasco?" *Economy and Environment,* fall 1992, pp. 1–2.
37. Ibid.
38. Thomas Berry, *The Dream of the Earth* (San Francisco: Sierra Club Books, 1988), p. 220.

12. After Brazil (pp. 179–190)

1. David E. Pitt, "Global Commission Sharpens Mission," *Earth Times,* 28 February 1993, p. 5.
2. "Global Environment Facility," *Bulletin and Quarterly Operational Summary,* November 1993, p. 1.
3. Mohammed El-Ashry, comments at GEF press briefing, 12 July 1994.
4. Nitin Desai, interview by author, September 1993.
5. "ICPD: UN Adopts Plan; Vatican Gives Limited Support," *Greenwire,* 14 September 1994, p. 3.
6. Nitin Desai, "Statement for Policy Coordination and Sustainable Development to the Commission on Sustainable Development," 16 May 1994, unpublished.
7. Peter Sand, "International Environmental Law after Rio," *European Journal of International Law,* 1993, p. 2.
8. Peter H. Sand, "UNCED and the Development of International Environmental Law," *Yearbook of International Environmental Law,* vol. 3, p. 22.
9. Sand, "International Environmental Law.
10. Maurice Strong, interview by author, October 1992.
11. David Buzzelli, interview by author, November 1993.
12. Frank Popoff, "Life after Rio: Merging Economics and Environmentalism" (speech to the Chemical Week conference, Washington, D.C., 15 October 1992).
13. "Iowa: State Looks to Adopt Lessons Learned in Rio," *Greenwire,* 30 March 1994, p. 11.
14. Maurice Strong, interview by author, October 1992.
15. Alicea Barcena, interview by author, September 1992.
16. Barbara Bramble, interview by author, November 1992.
17. Stong, interview, October 1992.
18. "Climate Change: EU Carbon Tax Shelved," *Greenwire,* 19 December 1994, p. 11.
19. "Biodiversity: UNEP Selected as Convention Secretariat," *Greenwire,* 14 December 1994.
20. John Stackhouse, "Rio Simmit Little More Than a Memory," *Toronto Globe and Mail,* 10 December 1994.
21. "Tokyo Declaration 1994," Tokyo Conference on Global Environmental Action, 26 October 1994, p. 6.
22. "U.N. Head Announces New Agenda for Development," *Terraviva,* 15 November 1994, p. 4.
23. Stackhouse, "Rio Summit."
24. Background interview by author, September 1994.
25. Ibid.
26. Barbara Crossette, "The Third World Is Dead, but Spirits Linger," *New York Times,* 13 November 199?, "News of the Week in Review."
27. Brian Urquhart, interview by author, March 1992.
28. *Greenwire,* 2 December 1994.

29. Maurice F. Strong, "Beyond Rio: New World Order or Lost Opportunity?" Environment Lecture at the Royal Botanic Garden, Kew, U.K., 16 September 1993.

13. Doing It (pp. 194–207)

1. Hanspeter Liniger, interview by author, May 1993.
2. Joseph Lynam, interview by author, May 1993.
3. Roger D. Stone, *The Nature of Development* (New York: Alfred A. Knopf, 1992), pp. 190–91.
4. Gilbert Arum, interview by author, May 1993.
5. Kagiso P. Keatimilwe, interview by author, May 1993.
6. Gaogakwe Phorano, interview by author, May 1993.
7. Robert Monro, interview by author, May 1993.
8. George N. Pangeti, deputy director, Department of National Parks and Wild Life Management, interview by author, May 1993.
9. Simon Metcalfe, "The Zimbabwe Communal Areas Management Program for Indigenous Resources (CAMPFIRE)" (paper presented at the Liz Claiborne–Art Ortenberg Foundation Community Based Conservation Workshop, Airlie, Va., 18–22 October 1993, pp. 35–44.
10. Monro, interview.
11. James Kamara, interview by author, May 1993.
12. E. T. Mundangepfupfu, interview by author, May 1993.
13. Philip Shabecoff, "The Working Rainforest," *Destination Discovery,* October 1993, p. 10.
14. "Bottom/Up Strategies Developed for Siberia's Lake Baikal Region," *International Society for Ecological Economics Newsletter,* April 1993, p. 7.
15. W. Wayt Gibbs, "No-Polluting Zone, Russia Follows Adirondack Approach to Environmental Protection," *Scientific American,* December 1994, p. 14.
16. Susan Witt, interview by author, October 1994.
17. "Bottom/Up Strategies."
18. The information on Molokai was obtained for a reporting trip in August 1994 to the island sponsored by the Nature Conservancy and the East-West Center at the University of Hawaii for U.S. and Japanese environmental journalists, in which the author participated.
19. Patrick Johnston, "Molokai Activist Plants Seeds for Microentrerprise Development," *Ka Wai Ola O Oha,* n.d.
20. Yolanda Rivera, "Once upon a Time . . . Banana Kelly and the Bronx Community Paper Company," *Banana Kelly CIA,* n.d. p. 1.
21. Weld F. Royal, "Paper Mill Project Plants Roots in the South Bronx," *BioCycle,* July 1994, p. 48.
22. John Holusha, "Pioneering Bronx Plant to Recycle City's Paper," *New York Times,* 6 May 1994.
23. "De-inking the Urban Forest," *New Partnerships in the Americas* (New Partnerships Working Group), December 1994, p. 40.

24. Allen Hershkowitz, interview by author, November 1994.
25. David Western and Michael Wright, "Issues in Community Based Conservation" (paper presented at the Liz Claiborne–Art Ortenberg Foundation Community Based Conservation Workshop, Airlie, Va., 18–22 October 1992).
26. Stephen Viederman, "The Economics of Sustainability: Challenges" (paper presented at workshop, the Economics of Sustainability, of the Fundacão Joaquim Mabuco, Recife, Brazil, 13–15 September 1994.
27. Fabio Feldmann, "National Policy Context for Community Action," paper presented at the Liz Claiborne–Art Ortenberg Foundation Community Based Conservation Workshop, Airlie, Va., 18–22 October 1992.
28. Viederman, "Economics of Sustainability."
29. John G. Robinson, "The Limits to Caring: Sustainable Living and the Loss of Biodiversity," *Conservation Biology,* 7(1) (March 1993): 20.
30. Lester R. Brown, Christopher Flavin, and Sandra Postel, *Saving the Planet* (New York: W. W. Norton, 1991), p. 21.

14. June 2042 (pp. 209–220)

1. Norman Myers and Julian L. Simon, *Scarcity or Abundance* (New York: W. W. Norton, 1994), p. 6.
2. M. I. Budyko, *Climactic Changes* (American Geophysical Union, 1977), quoted in Barbara K. Rodes and Rice Odell, *A Dictionary of Environmental Quotations* (New York: Simon & Schuster, 1992), p. 112.
3. Eric Ashby, *Reconciling Man with Environment* (1978) quoted in Rodes and Odell, *Dictionary of Enviromental Quotations*, p. 112.
4. Gwyn Prins, *Threats without Enemies* (London: Earthscan Publications, 1993), p. 2.
5. Quoted in Robert McHenry and Charles Van Doren, eds., *A Documentary History of Conservation in America* (New York: Praeger, 1972), p. 172.
6. Robert D. Kaplan, "The Coming Anarchy," *Atlantic Monthly,* February 1984, p. 46.
7. Leon F. Bouvier and Lindsey Grant, *How Many Americans?* (San Francisco: Sierra Club Books, 1994), p. 95.
8. Harlan Cleveland, "The Management of Peace," *GAO Journal,* winter 1990/91, p. 11.
9. Anthony D. Cortese, "Earth Day 1995: Partnerships for Sustainability" (paper presented to the New England Earth Day Organizing Conference, John F. Kennedy School of Government, Cambridge, Mass., 5 November 1994).
10. Lionel Tiger and Robin Fox, *The Imperial Animal* (New York: Holt, Rinehart and Winston, 1971), p. 211.
11. Fernand Braudel, *The Perspective of the World,* vol. 3 of *Civilization and Capitalism* (New York: Harper & Row, 1971), p. 628.
12. Ibid., p. 632.
13. John Passmore, *Man's Responsibility for Nature* (New York: Charles Scribner's Sons, 1974), p. 188.

14. Ralph Buultjens, "Reaching Out in the Post-Rio World," *Earth Times,* 28 February 1993, p. 14.
15. Vaclav Havel, "The End of the Modern Era" (excerpts from a speech presented at the World Economics Forum in Davos, Switzerland, 4 February 1992), *New York Times,* 1 March 1992, "News of the Week in Review."
16. Lynton Keith Caldwell, *Between Two Worlds: Science, the Environmental Movement and Policy Choice* (Cambridge: Cambridge University Press, 1990), p. 196.
17. Boutros Boutros-Ghali, "Beleaguered Are the Peacekeepers," *New York Times.*
18. Franklin L. Baumer, *Modern European Thought* (New York: Macmillan, 1977), p. 415.
19. Ferenc Maté, *A Reasonable Life: Toward a Simpler, Secure, More Humane Existence* (New York: W. W. Norton, 1993).
20. James A. Lovelock, *Gaia: A New Look at Life*, 1979, quoted in Rodes and Odell, *Dictionary of Environmental Quotations,* p. 113.
21. Herman E. Daly and John B. Cobb Jr., *For the Common Good* (Boston: Beacon Press, 1989), p. 400.

Selected Bibliography

Barnet, Richard J. *The Lean Years.* New York: Simon & Schuster, 1980.

Barnet, Richard J., and John Cavanagh. *Global Dreams.* New York: Simon & Schuster, 1994.

Benedick, Richard Elliot. *Ozone Diplomacy.* Cambridge, Mass.: Harvard University Press, 1991.

Berry, Thomas. *The Dream of the Earth.* San Francisco: Sierra Club Books, 1988.

Boardman, Robert. *International Organization and the Conservation of Nature.* Bloomington: Indiana University Press, 1981.

Bormann, F. Hebert, and Stephen R. Kellert, eds. *Ecology, Economics, Ethics: The Broken Circle.* New Haven, Conn.: Yale University Press, 1991.

Bouvier, Leon F., and Lindsey Grant. *How Many Americans?* San Francisco: Sierra Club Books, 1994.

Bramwell, Anna. *Ecology in the 20th Century.* New Haven, Conn.: Yale University Press, 1989.

Braudel, Fernand. *The Perspective of the World.* Vol. 3 of *Civilization and Capitalism.* New York: Harper & Row, 1981.

———. *The Structures of Everyday Life.* Vol. 1 of *Civilization and Capitalism.* New York: Harper & Row, 1981.

Brown, Edith Weiss. "Environment and Trade as Partners in Sustainable Development: A Commentary." *American Journal of International Law* 86 (October 1992): 728.

———. "International Environmental Law: Contemporary Issues and the Emergence of a New World Order." *Georgetown Law Journal* 81(3) (March 1993), 165.

Brown, Harrison. *The Challenge of Man's Future.* New York: Viking Press, 1954.

Brown, Lester. *Building a Sustainable Society.* New York: W. W. Norton, 1981.

Brown, Lester, Hal Kane, Ed Ayres. *Vital Signs 1993.* New York: W. W. Norton, 1993.

Brown, Lester, et al., eds. *State of the World 1990.* New York: W. W. Norton, 1990.

———. *State of the World 1994.* New York: W. W. Norton, 1994.

Caldicott, Helen, M.D. *If You Love This Planet.* New York: W. W. Norton, 1992.

Caldwell, Lynton Keith. *Between Two Worlds: Science, the Environmental Movement and Policy Choice.* Cambridge: Cambridge University Press, 1990.

———. "Globalizing Environmentalism: Threshold of a New Phase in International Relations." School of Public and Environmental Affairs, Indiana University, Bloomington, 10 August 1990.

———. *International Environmental Policy,* 2d ed. Durham, N.C.: Duke University Press, 1990.

Caring for the Earth: A Strategy for Sustainable Living. Gland, Switzerland: International Union for the Conservation of Nature, United Nations Environment Programme, World Wildlife Fund, 1991.

Carson, Rachel. *Silent Spring.* 25th anniversary edition. Boston: Houghton Mifflin, 1987.

Challenge to the South: An Overview and Summary of the South Commission Report. Geneva: The South Commission, 1990.

Chivian, Eric, Michael McCally, Howard Hu, and Andrew Haines. *Critical Condition: Human Health and the Environment.* Cambridge, Mass.: M.I.T. Press, 1993.

Choosing a Sustainable Future: The Report of the National Commission on the Environment. Washington, D.C.: Island Press, 1993.

Choucri, Nazli, ed. *Global Accord.* Cambridge, Mass.: M.I.T. Press, 1993.

Cleveland, Harlan. "The Management of Peace." *GAO Journal,* winter 1990): 11.

Commoner, Barry. *The Closing Circle.* New York: Alfred A. Knopf, 1971.

———. *Making Peace with the Planet.* New York: Pantheon Books, 1990.

Costanza, Robert, ed. *Ecological Economics: The Science and Management of Sustainability.* New York: Columbia University Press, 1991.

Crump, Andy. *Dictionary of Environment and Development.* Cambridge, Mass.: M.I.T. Press, 1993.

Daly, Herman E. "Boundless Bull." *Gannett Center Journal,* summer 1990.

———. *Steady State Economics.* 2d ed. Washington, D.C.: 1991.

Daly, Herman E., and Kenneth N. Townsend, eds. *Valuing the Earth.* Cambridge, Mass.: M.I.T. Press, 1993.

Daly, Herman E., and John B. Cobb Jr. *For the Common Good.* Boston: Beacon Press, 1989.

Defending the Earth: Abuses of Human Rights and the Environment. New York: Human Rights Watch of the Natural Resources Defense Council, 1992.

De Onis, Juan. *The Green Cathedral.* New York: Oxford University Press, 1992.

Dunlap, Riley E., George H. Gallup Jr., and Alec M. Gallup. "Of Global Concern: Results of the Health of the Planet Survey," *Environment* 35(9) (November 1993).

Durning, Alan. *Action at the Grassroots: Fighting Poverty and Environmental Decline.* Worldwatch Paper 88. Washington, D.C.: Worldwatch Institute, 1989.

———. *How Much Is Enough?* New York: W. W . Norton, 1992.

Eckholm, Erik P. *Down to Earth.* New York: W. W. Norton, 1982.

Ehrlich, Paul R. *The Population Bomb.* Rivercity, Mass.: Rivercity Press, 1975.

Ehrlich, Paul R., and Anne H. Ehrlich. *Healing the Planet.* Reading, Mass.: Addison-Wesley, 1991.

Flynn, John. "The Subversive Scientists." *Amicus Journal,* fall 1990.

Funtowicz, Silvio O., and Jerome R. Ravetz. "A New Scientific Methodology for Global Environment Issues." In *Ecological Economics,* edited by Robert Costanza. New York: Columbia University Press, 1991.

Glacken, Clarence. *Traces on the Rhodian Shore.* Berkeley and Los Angeles: University of California Press, 1967.

Glendenning, Chellis. *When Technology Wounds.* New York: William Morris, 1990.

The Global Partnership for Environment and Development: A Guide to Agenda 21. Geneva: UN Conference on Environment and Development, 1992.

Global Tomorrow Coalition. *The Global Ecology Handbook.* Boston: Beacon Press, 1990.

Goldsmith, Edward. *The Way*. Boston: Shambala, 1993.
Gore, Al. *Earth in the Balance*. Boston: Houghton Mifflin, 1992.
Griffin, Susan. *Woman and Nature*. New York: Harper Colophon Books, 1978.
Haas, Peter M., Robert O. Koehane, and Marc A. Levy, eds. *Institutions for the Earth*. Cambridge, Mass.: M.I.T. Press, 1993.
Haas, Peter M., Marc A. Levy, and Edward A. Parson. "Appraising the Earth Summit." *Environment,* October 1992.
Henderson, Hazel. *The Politics of the Solar Age: Alternatives to Economics*. Indianapolis, Ind.: Knowledge Systems, 1988.
Homer-Dixon, Thomas F. *Environmental Change and Violent Conflict*. Occasional Paper No. 4. New York: American Academy of Arts and Sciences, 1990.
Huth, Hans. *Nature and the American*. Lincoln: University of Nebraska Press, 1990.
Idris, S. M. Mohamed. "What Green Really Means for the Third World." *Third World Resurgence* 1(1) (Spring 1990).
Jacobson, Jodi L. *Gender Bias: Roadblock to Sustainable Development*. Worldwatch Paper 100. Washington, D.C.: Worldwatch Institute, 1992.
Jamison, Andrew, Ron Eyerman, Jacqueline Cramer. *The Making of the New Environmental Consciousness*. Edinburgh: Edinburgh University Press, 1990.
Jancar, Barbara. "Chaos as an Explanation of the Role of Environmental Groups in East European Politics." In *Green Politics Two,* edited by Wolfgang Rudig. Edinburgh: Edinburgh University Press, 1991.
Jhamtani, Hira. "The Imperialism of Northern NGOs." *Earth Island Journal,* summer 1992.
Kane, Michael J. *International Protection of Human Rights and the Environment*. Washington, D.C.: U.S. Environmental Protection Agency, 1991.
Kaplan, Robert D. "The Coming Anarchy." *Atlantic Monthly,* February 1984.
Kennedy, Paul. *Preparing for the Twenty-First Century*. New York: Random House, 1993.
Kissinger, Henry. *Diplomacy*. New York: Simon & Schuster, 1994.
Lee, Kai N. *Compass and Gyroscope*. Washington, D.C.: Island Press, 1993.
Leopold, Aldo. *A Sand County Almanac*. New York: Oxford University Press, 1943.
MacNeill, Jim, Peter Winsemius, and Taizo Yakushiji. *Beyond Interdependence: The Meshing of the World's Economy and the Earth's Ecology*. New York: Oxford University Press, 1991.
Marsh, George Perkins. *Man and Nature*. Edited by David Lowenthal. Cambridge, Mass.: Harvard University Press, 1965.
Martinez-Alier, Juan. *Ecological Economics*. Oxford: Basil Blackwell, 1987.
Mason, Stephen F. *A History of the Sciences*. New York: Collier Books, 1962.
Maté, Ferenc. *A Reasonable Life: Toward a Simpler, Secure, More Humane Existence*. New York: W. W. Norton, 1993.
Mathews, Jessica Tuchman, ed. *Preserving the Global Environment*. New York: W. W. Norton, 1991.

McCormick, John. *Reclaiming Paradise: The Global Environmental Movement.* Bloomington: Indiana University Press, 1991.

McHenry, Robert, with Charles Van Doren, eds. *A Documentary History of Conservation in America.* New York: Crown, 1972.

McNamara, Robert S. *A Global Population Policy to Advance Human Development.* New York: United Nations, 1991.

Meadows, Donella H., Dennis L. Meadows, and Jörgen Randers. *Beyond the Limits.* Post Mills, Vt.: Chelsea Green, 1992.

Meadows, Donella H., Dennis L. Meadows, Jörgen Randers, and William W. Behrens III. *The Limits to Growth.* New York: Universe Books, 1972.

Mewes, Horst. "The Green Party Comes of Age." *Environment* 27(5) (June 1985).

Michaels, Patrick J. *Sound and Fury.* Washington, D.C.: Cato Institute, 1992.

Mikesell, Raymond F., and Lawrence F. Williams. *International Banks and the Environment.* San Francisco: Sierra Club Books, 1992.

Milbrath, Lester W. *Environmentalists: Vanguard for a New Society.* Albany: State University of New York Press, 1984.

Models of Sustainable Development: Exclusive or Complementary Approaches of Sustainability. 2 vols. Papers presented at an International Symposium at the Université Pantheon-Sorbonne, 16–18 March, 199. Paris: Association Française des Sciences et Technologies de L'Information et des systèmes (AFCET), 1994.

Moore, Curtis, and Alan Miller. *Green Gold.* Boston: Beacon Press, 1994.

Monrgenstern, Richard D., and Dennis Tirpak. "Projecting the Impacts of Global Warming." *EPA Journal* 15(1), (January/February 1989).

Mumford, Lewis. *Technics and Civilization.* New York: Harcourt, Brace & World, 1962.

Myers, Norman. *Ultimate Security.* New York: W. W. Norton, 1993.

Myers, Norman, and Julian L. Simon. *Scarcity or Abundance.* New York: W. W. Norton, 1994.

Nash, Roderick Frazier. *The Rights of Nature.* Madison: University of Wisconsin Press, 1989.

Oppenheimer, Michael, and Robert H. Boyle. *Dead Heat.* New York: Basic Books, 1990.

Paehlke, Robert C. *Envionrmentalism and the Future of Progressive Politics.* New Haven, Conn.: Yale University Press, 1989.

Passmore, John. *Man's Responsibility for Nature.* New York: Charles Scribner's Sons, 1974.

Pearce, David W. *Economic Values and the Natural World.* Cambridge, Mass.: M.I.T. Press, 1993.

Pearce, David, Anil Markandy, and Edward B. Barbier. *Blueprint for a Green Economy.* London: Earthscan International, 1989.

Piel, Gerald, and Osborn Segerberg Jr., eds. *The World of René Dubos.* New York: Henry Holt, 1990.

Ponting, Clive. *A Green History of the World: The Environment and the Collapse of Great Civilizations.* New York: Penguin Books, 1991.

Pool, Robert. "Struggling to Do Science for Society." *Science* 248 (May 1990).
Porritt, Jonathan. *Seeing Green.* Oxford: Basil Blackford, 1984.
Porter, Gareth, and Janet Welsh Brown. *Global Environmental Politics.* Boulder, Colo.: Westview Press, 1991.
Postel, Sandra. *Defusing the Toxics Threat.* Worldwatch Paper 79. Washington, D.C.: Worldwatch Institute, 1987.
Pryns, Gwyn, ed. *Threats without Enemies: Facing Environmental Insecurity.* London: Earthscan Publications, 1993.
Ramphal, Sir Shridath. *Our Country the Planet.* Washington, D.C.: Island Press, 1992.
Rifkin, Jeremy. *Biosphere Politics.* New York: Crown, 1991.
Robinson, John G. "The Limits to Caring: Sustainable Living and the Loss of Biodiversity." *Conservation Biology* 7(1) (March 1993).
Robinson, Nicholas. "Diplomacy's New Frontiers." *Gale Environmental Almanac.* Detroit: Gale Research, 1993.
Rodes, Barbara K., and Rice Odell. *A Dictionary of Environmental Quotations.* New York: Simon & Schuster, 1992.
Roszak, Theodore. *The Voice of the Earth.* New York: Simon & Schuster, 1992.
Rousseau, Jean Jacques. *On the Inequality among Mankind.* Vol. 24 of the Harvard Classics. New York: P. F. Collier & Son, 1909.
Rudig, Wolfgang. "Green Party Politics around the World." *Environment.* October 1991.
Russell, Milton. "Environmental Protection for the 1990s and Beyond." *Environment* 29(7) (September 1987).
Sand, Peter H. "International Environmental Law after Rio." *European Journal of International Law,* 1993.
———. *Lessons Learned in Global Governance.* Washington, D.C.: World Resources Institute, 1990.
Schmidheiny, Stephan, with the Business Council on Sustainable Development. *Changing Course.* Cambridge, Mass.: M.I.T. Press, 1992.
Schneider, Stephen H. *Global Warming.* San Francisco: Sierra Club Books, 1989.
Schumacher, E. F. *Small Is Beautiful: Economics As If People Mattered.* New York: Harper & Row, 1973.
Shabecoff, Philip. *A Fierce Green Fire: The American Environmental Movement.* New York: Hill and Wang, 1993.
Silver, Cheryl Simon, with Ruth S. DeFries. *One Earth One Future.* Washington, D.C.: National Academy Press, 1990.
Simon, Julian, and Herman Kahn. *The Resourceful Earth.* New York: Basil Blackwell, 1984.
Sohn, Louis B. "The Stockholm Declaration on the Human Environment." *Harvard International Law Journal* 14(3) (summer 1973).
Spretnak, Charlene, and Fritjof Capra. *Green Politics.* Santa Fe, N.M.: Bear & Co., 1986.
Starke, Linda. *Signs of Hope.* New York: Oxford University Press, 1990.
Stone, Christopher D. *The Gnat Is Older than Man.* Princeton, N.J.: Princeton University Press, 1992.

Stone, Peter. *Did We Save the Earth at Stockholm?* London: Earth Island, 1973.

Stone, Roger D. *The Nature of Development.* New York: Alfred A. Knopf, 1992.

Stone, Roger D., and Eve Hamilton. *Global Economics and the Environment.* New York: Council on Foreign Relations Press, 1991.

Strong, Maurice F. *From Stockholm to Rio: A Journey Down a Generation.* Earth Summit Publication no. 1. New York: United Nations Conference on Environment and Development, 1991.

———. "One Year after Stockholm: An Ecological Approach to Management." *Foreign Affairs* (July 1973).

Susskind, Lawrence E. *Environmental Diplomacy.* New York: Oxford University Press, 1994.

Swift, Adam. *Global Political Ecology.* London: Pluto Press, 1993.

Tappa, Louise. "The Theology of Environment and Development." In *SONED on UNCED: A Southern Perspective on the Environment and Development Crisis.* Geneva: Southern Networks for Development, African Region, 1991.

Thomas, Caroline. *The Environment in International Relations.* London: Royal Institute of International Affairs, 1992.

Thoreau, Henry David. *Walden.* London: J. M. Dent & Sons, 1930.

Trends in International Environmental Law. By the editors of the *Harvard Law Review.* Chicago: American Bar Association, Section of International Law and Practice, 1992.

Turner, B. L. II, William C. Clark, Robert W. Kates, John F. Richards, Jessica T. Mathews, and William B. Meyer, eds. *The Earth as Transformed by Human Action.* Cambridge: Cambridge University Press with Clark University, 1990.

United Nations Population Fund. *Population Pressures: A Complex Equation.* New York: United Nations, 1991.

Urquhart, Brian. *A Life in Peace and War.* New York: Harper & Row, 1987.

Vig, Norman J., and Michael E. Kraft, eds. *Environmental Policy in the 1990s.* Washington, D.C.: CQ Press, 1990.

Vitousek, P. M., P. R. Ehrlich, A. H. Ehrlich, and P. M. Matson. "Human Appropriation of the Products of Photsynthesis." *Bioscience* 36(6) (1986).

Von Weizacker, Ernst U., and Jochen Jesinghaus. *Ecological Tax Reform.* London: Zed Books, 1993.

Ward, Barbara, and René Dubos. *Only One Earth.* New York: W. W. Norton, 1972.

Wertenbaker, William. "A Reporter at Large: The Law of the Sea." *New Yorker,* August 1, 1983.

White, Lynn Jr. "The Historical Roots of Our Ecological Crisis," *Science* 155 (March 1967).

Willums, Jan-Olaf, and Ulrich Goluke. *From Ideas to Action: Business and Sustainable Development.* Oslo: International Environment Bureau of the International Chamber of Commerce, 1992.

Wilson, Edward O. *The Diversity of Life.* Cambridge, Mass.: Belknap Press of Harvard University Press, 1992.

World Commission on Environment and Development. *Our Common Future.* New York: Oxford University Press, 1987.

World Resources Institute. *World Resources 1990–91.* New York: Oxford University Press, 1990.
Worster, Donald. *Nature's Economy.* Cambridge: Cambridge University Press, 1977.
Young, John. *Sustaining the Earth.* Cambridge, Mass.: Harvard University Press, 1990.

Index

University Press of New England publishes books under its own imprint and is the publisher for Brandeis University Press, Dartmouth College, Middlebury College Press, University of New Hampshire,University of Rhode Island, Tufts University, University of Vermont, Wesleyan University Press, and Salzburg Seminar.

Library of Congress Cataloging-in-Publication Data

Shabecoff, Philip.
A new name for peace : international environmentalism, sustainable development, and democracy / Philip Shabecoff.
p. cm.
Includes bibliographical references and index.
ISBN 0-87451-688-9 (cl : alk. paper)
1. Environmentalism. 2. Sustainable development. 3. Geopolitics.
I. Title.
GE195.S43 1996
363.7—dc20 96-2235
♾